NP1L1360   PA76232

The Nottingham Trent University
Library & Information Services

2 APR 1998

2 1 MAY 1998

WITHDRAWN

Boots Library, Goldsmith Street,
Nottingham, NG1 5LS.
Telephone: (0115) 948 6434

Please return the book by the last date stamped.
Fines will accrue on overdue books, but renewals may be
requested in person, by phone or by post.

Boots Date Label, 98

Telephone 486434

 Dryden Street, Nottingham, NG1 4FZ
Telephone 48248   Ext. 2363

40 0116333 3

IEE Power Engineering Series 6

Series Editors: Dr A.T. Johns
                  G. Ratcliff
                  Prof. A. Wright

HIGH VOLTAGE DIRECT CURRENT TRANSMISSION

J. Arrillaga MSc.Tech., Ph.D., D.Sc., C.Eng., F.I.E.E.
Professor of Electrical Engineering, University of Canterbury,
New Zealand.

Peter Peregrinus Ltd.
On behalf of the Institution of
Electrical Engineers

Previous volumes in this series

Volume 1	Power circuit breaker theory and design. C.H. Flurscheim (Editor)
Volume 2	Electric fuses A. Wright & P.G. Newbery
Volume 3	Z-transform electromagnetic transient analysis in high-voltage networks W. Derek Humpage
Volume 4	Industrial microwave heating A.C. Metaxas and R.J. Meredith
Volume 5	Power system economics T.W. Berrie

Published by: Peter Peregrinus Ltd., London, UK.

© 1983: Peter Peregrinus Ltd.

All rights reserved. No part of this publication may be reproduced, stored in a retrieval system or transmitted in any form or by any means—electronic, mechanical, photocopying, recording or otherwise—without the prior written permission of the publisher.

British Library Cataloguing in Publication Data

Arrillaga, J.
   High voltage direct current transmission.
   —(IEE power engineering series; 6)
   1. Electric power distribution—Direct current
   2. Electric power distribution—High tension
   I. Title   II. Series
   621.319'12   TK3111

ISBN 0 906048 97 4

Printed in England by Short Run Press Ltd., Exeter

Contents

Preface ix

1 Development of a h.v.d.c. technology 1
   1.1 Introduction 1
   1.2 Brief historical background 1
   1.3 The mercury-arc valve 3
   1.4 Mercury-arc schemes 5
       1.4.1 Sweden–Gotland link (1954) 5
       1.4.2 English Channel (1961) 5
       1.4.3 Volgograd–Donbas (1962–65) 5
       1.4.4 New Zealand link (1965) 5
       1.4.5 Konti–Skan (1965) 6
       1.4.6 Sakuma interconnection (1965) 6
       1.4.7 Sardinia–Italy (mainland) (1967) 6
       1.4.8 Pacific intertie (1970) 6
       1.4.9 Kingsnorth scheme (1974) 6
       1.4.10 Nelson River Bipole 1 (1973–77) 7
   1.5 The solid state technology 7
   1.6 Thyristor take-over 9
       1.6.1 Eel River (1972) 9
       1.6.2 Cabora–Bassa (1977–79) 9
       1.6.3 Inga–Shaba (1981) 10
       1.6.4 Skagerrak (1976–77) 10
       1.6.5 Square Butte (1977) 10
       1.6.6 Nelson River Bipole 2 (1978–85) 11
       1.6.7 Itaipu scheme 11
       1.6.8 Ekibastuz–Centre 11
       1.6.9 New cross-channel link 11
   1.7 Operation reliability 12
   1.8 Future expansion 12
   1.9 References 13

2 Static power conversion 14
   2.1 Introduction 14
   2.2 Basic conversion principle 14
   2.3 Selection of convertor configuration 16

2.4	The ideal commutation process		16
	2.4.1 Effect of gate control		18
	2.4.2 Valve current and voltage waveforms		18
2.5	The real commutation process		22
	2.5.1 Commutating voltage		22
	2.5.2 Commutation reactance		22
	2.5.3 Analysis of the commutation circuit		24
2.6	Rectifier operation		28
	2.6.1 Mean direct voltage		30
	2.6.2 A.C. current		30
2.7	Invertor operation		31
2.8	Power factor and reactive power		33
2.9	Convertor harmonics		35
	2.9.1 Characteristic harmonics		37
	2.9.2 Non-characteristic harmonics		42
2.10	References		50

3 Harmonic elimination — 51
- 3.1 Introduction — 51
- 3.2 Pulse number increase — 51
- 3.3 Design of a.c. filters — 52
  - 3.3.1 Design criteria — 52
  - 3.3.2 Design factors — 53
  - 3.3.3 Network impedance — 55
  - 3.3.4 Circuit modelling — 58
  - 3.3.5 Tuned filters — 59
  - 3.3.6 Self-tuned filters — 63
  - 3.3.7 High pass filters — 63
  - 3.3.8 Example of recent filter arrangement — 64
  - 3.3.9 Type C damped filters — 64
  - 3.3.10 Simplified filtering for 12-pulse convertors — 65
- 3.4 D.C. side filters — 66
- 3.5 Alternative methods of harmonic elimination — 68
  - 3.5.1 Magnetic flux compensation — 68
  - 3.5.2 Harmonic injection — 70
  - 3.5.3 D.C. ripple injection — 72
- 3.6 References — 75

4 Control of h.v.d.c. convertors and systems — 76

A Convertor control — 76

- 4.1 Basic philosophy — 76
- 4.2 Individual phase-control — 77
- 4.3 Equidistant firing control — 79
  - 4.3.1 Constant current loop — 80
  - 4.3.2 Invertor extinction angle control — 81
  - 4.3.3 Transition from extinction angle to current control — 82
  - 4.3.4 Other equidistant firing control schemes — 82
  - 4.3.5 Application to 12-pulse convertor groups — 86
- 4.4 Comparative merits — 86
- 4.5 Analogue and digital controls — 87

	B	D.C. system control	88
	4.6	Basic philosophy	88
	4.7	Characteristics and direction of d.c. power flow	88
		4.7.1 Reversal of power flow	92
		4.7.2 Modifications to the basic characteristics	93
		4.7.3 Tap changer control	94
		4.7.4 Different control levels	95
		4.7.5 Power flow control	96
		4.7.6 Telecommunication requirements	97
	4.8	References	98
5	**Interaction between a.c. and d.c. systems**		**99**
	5.1	Introduction and definitions	99
	5.2	Voltage interaction	102
		5.2.1 Dynamic voltage regulation	103
		5.2.2 Dynamic compensation	105
	5.3	Harmonic instabilities	106
		5.3.1 Instabilities caused by individual firing control	106
		5.3.2 Convertor transformer saturation effects	109
		5.3.3 Core saturation instability	112
		5.3.4 Generalisation of the instability problem	114
	5.4	D.C. power modulation	115
		5.4.1 Frequency control	115
		5.4.2 Power/frequency control	116
		5.4.3 Dynamic stabilisation of a.c. systems	116
		5.4.4 Large signal modulation	117
		5.4.5 Controlled damping of d.c.-interconnected systems	118
		5.4.6 Damping of subsynchronous resonances	120
		5.4.7 Active and reactive power coordination	121
		5.4.8 Overall control coordination	122
		5.4.9 Transient stabilisation of a.c. systems	124
	5.5	References	125
6	**Main design considerations**		**126**
	6.1	Introduction	126
	6.2	Thyristor convertors	127
		6.2.1 Thyristor valve architecture	127
		6.2.2 Twelve-pulse convertor unit	127
		6.2.3 Multibridge convertors	130
		6.2.4 Valve cooling systems	134
		6.2.5 Valve control circuitry	134
		6.2.6 Valve protective functions	135
		6.2.7 Thyristor valve tests	136
		6.2.8 Convertor circuits and components	137
		6.2.9 Thyristor station layout	140
		6.2.10 Relative costs of convertor components	143
	6.3	Mercury-arc circuit components	143
		6.3.1 Valve group	143
		6.3.2 Convertor station	147
		6.3.3 Mercury-arc convertor layout	147
	6.4	Convertor transformers	148
	6.5	Smoothing reactors	149

Contents

6.6	Overhead lines	149
6.7	Cable transmission	151
6.8	Earth electrodes	152
6.9	Design of back-to-back thyristor convertor systems	154
6.10	References	156

7 Fault development and protection — **158**
- 7.1 Introduction — 158
- 7.2 Convertor disturbances — 158
  - 7.2.1 Misfire and firethrough — 159
  - 7.2.2 Commutation failure — 159
  - 7.2.3 Backfire — 162
  - 7.2.4 Internal short-circuits — 164
  - 7.2.5 Bypass action — 164
  - 7.2.6 Bypass action in thyristor bridges — 165
- 7.3 Simulation of practical disturbances — 167
- 7.4 A.C. system faults — 170
  - 7.4.1 Three-phase faults — 170
  - 7.4.2 Unsymmetrical faults — 172
- 7.5 D.C. line fault development — 173
  - 7.5.1 Fault detection — 173
  - 7.5.2 Fault clearing and recovery — 174
  - 7.5.3 Overall dynamic response — 175
- 7.6 Overcurrent protection — 175
  - 7.6.1 Valve group protection — 177
  - 7.6.2 D.C. line protection — 181
  - 7.6.3 Filter protection — 182
- 7.7 References — 182

8 Transient overvoltages and insulation coordination — **183**
- 8.1 Introduction — 183
- 8.2 Overvoltages excited by disturbances on the d.c. side — 185
- 8.3 Harmonic overvoltages excited by a.c. disturbances — 186
- 8.4 Overvoltages due to convertor disturbances — 188
- 8.5 Fast transients generated on the d.c. system — 188
  - 8.5.1 Lightning surges — 189
  - 8.5.2 Switching-type surges — 189
- 8.6 Surges generated on the a.c. system — 192
- 8.7 Fast transient phenomena associated with the convertor plant — 193
  - 8.7.1 Mercury-arc convertors — 193
  - 8.7.2 Thyristor convertors — 196
- 8.8 Insulation coordination — 198
  - 8.8.1 System design — 199
  - 8.8.2 Surge arresters — 199
  - 8.8.3 Application of surge arresters — 200
- 8.9 References — 204

9 D.C. versus a.c. transmission — **206**
- 9.1 General considerations — 206
- 9.2 Bulk energy transfer — 208
  - 9.2.1 A comparison of a.c. and d.c. transmission characteristics — 208
  - 9.2.2 Power-carrying capability of a.c. and d.c. lines — 210
  - 9.2.3 Equivalent reliability criterion — 212

		9.2.4	Effect of losses and discount rates	213
		9.2.5	Other considerations	213
		9.2.6	Infeeds at lower voltage levels	215
		9.2.7	Environmental effects	216
	9.3	System interconnection		218
	9.4	References		219
10	**Research and development**			**220**
	10.1	Introduction		220
	10.2	D.C. circuit breakers		220
		10.2.1	Use of h.v.d.c. circuit breakers in point-to-point interconnections	222
	10.3	Multiterminal d.c. transmission		224
		10.3.1	Technical comparisons	224
		10.3.2	Economic comparisons	226
		10.3.3	Fault detection	227
		10.3.4	Switching requirements	227
	10.4	Generator–Rectifier Units		229
		10.4.1	Unit connection using controlled rectifiers	229
		10.4.2	Unit connection using diode rectifiers	231
	10.5	Forced commutation		232
	10.6	Existing a.c. transmission facilities converted for use with d.c.		233
	10.7	Compact convertor stations		234
	10.8	Microprocessor-based digital control		237
	10.9	General conclusion		240
	10.10	References		240

Subject index   **243**

Preface

Following the successful completion of the first commercial application of h.v.d.c. transmission in 1954, two of my former colleages, C. Adamson and N. G. Hingorani, collected most of the information available at the time in a book entitled 'High Voltage Direct Current Power Transmission', which appeared in 1960.

A decade later, one of the pioneers of the modern a.c. power system, the late E. W. Kimbark, completed his distinguished career by recording the state of the new technology in a book entitled 'Direct Current Transmission: Volume I'.

In 1975 another pioneer, this time of the modern d.c. power system, E. Uhlmann, recorded his vast experience in a comprehensive treatise, entitled 'Power Transmission by Direct Current', which included most valuable information, until then only available to a privileged few.

Almost another decade has passed since Dr. Uhlmann's effort and a century since the celebrated success of the Pearl Street d.c. power station. The last decade has seen the consolidation of the solid state technology and a wider acceptance of the fast controllability of h.v.d.c. convertors.

It seems therefore appropriate to report these advances in another book of perhaps more general interest, and I feel encouraged in such a task by Dr. Uhlmann's own words in the preface of his book "... and it is his hope that in the future there will be available a complete range of books covering all aspects of this field".

The book has been written with two objects in mind: From the educational point of view, h.v.d.c. transmission is an ideal subject to demonstrate the basic principles of static power conversion and the interaction taking place between large convertor plant and power systems. This book should therefore prove a valuable addition to the basic text books presently used for the teaching of power systems and power electronics in final year degree and postgraduate courses.

The second object is to meet the specific needs of the practicing engineers; in this respect the book covers the main developments which have made h.v.d.c. transmission a competitive technology.

I must acknowledge the valuable help received over the past twenty years from so many experts in the field. From my earlier days with the University of Manchester Institute of Science and Technology (UMIST), I am particularly grateful to C. Adamson, N. G. Hingorani, J. Reeve, P. C. S. Krishnayya, M. Morales, and especially to my wife Greta, who played a hidden but essential part in the dissemination of much of the h.v.d.c. material coming out of UMIST. Regular discussions with J. D. Ainsworth of GEC were always particularly helpful in moderating my enthusiastic academic views.

In more recent times, I would like to acknowledge the continued encouragement and support received from New Zealand Electricity which kept my interest in h.v.d.c. transmission alive, especially from P. W. Blakeley, K. D. McCool, K. S. Turner, M. D. Heffernan, B. J. Harker, M. T. O'Brien, P. S. Barnett and C. V. Currie. I am also endebted to my present research students for their critical comments during the preparation of the manuscript, in particular A. P. B. Joosten, J. C. Graham, and H. Hisha.

It would be difficult to properly acknowledge all the sources of information used in the preparation of the book; I must, however, single out the vast amount of work carried out by the CIGRE Committee 14 group on h.v.d.c. transmission, the many recent contributions by J. P. Bowles of Bodeven Inc. (Canada), and again the invaluable private communications from J. D. Ainsworth of GEC (Stafford, U.K.).

Finally, I wish to thank my secretary Mrs. A. Haughan for her active participation in the preparation of the manuscript.

Chapter 1
Development of a h.v.d.c. technology

1.1 Introduction

With the positive experience of many years of reliable h.v.d.c. system operation throughout the world, it is now possible to reiterate with confidence the claims more cautiously made in earlier references to the subject.

The main claims generally made in favour of the d.c. alternative are:

(a) D.C. transmission results in lower losses and costs than equivalent a.c. lines, but the terminal costs and losses are higher.
(b) A.C. transmission via cable is impractical over long distances. Such restriction does not exist with d.c.
(c) D.C. constitutes an asynchronous interconnection and does not raise the fault level appreciably.
(d) The power flow in a d.c. scheme can be easily controlled at high speed. Thus with appropriate controls, a d.c. link can be used to improve a.c. system stability.
(e) D.C. stations, with or without transmission distance, can be justified for the interconnection of a.c. systems of different frequencies or different control philosophies.

This book attempts to justify the above claims with reference to the present state of the art and particularly the thyristor technology.

1.2 Brief historical background

As early as 1881 Marcel Deprez, inspired by experiments of arc lights across a d.c. generator, published the first theoretical examination of h.v.d.c. power transmission. He soon put theory into practice and by 1882 he transmitted 1·5 kW at 2 kV over a distance of 35 miles.

The following decade witnessed the rising of alternating currents on account of the availability of transformers and the development of induction motors.

This prompted the following warning by Thomas Edison in 1887: "Take warning! Alternating currents are dangerous, they are fit only for the electric chair ...".

From 1889, R. Thury continued the work of Deprez by using d.c. generators in series to attain high transmission voltages. Among his many European installations, the best example of a d.c. transmission technology was that from Moutiers to Lyon with a final capacity of 20 MW at 125 kV over a distance of 230 km. This scheme operated at constant current and was used as a reinforcement of an existing a.c. system. It was probably the first recognition of a.c.–d.c. co-existence, as Thury himself put it "The two systems shake hands fraternally in order to give each other help and assistance ...".

With the advent of the steam turbine as a prime mover for the generation of power, the limitations of the Thury system were accentuated. There had been no special problems with the low speed water turbines driving the generators, but now the use of the steam turbine for d.c. generation depended on the availability of high-speed reduction gearing. However, by this time some other interesting developments had taken place.[1]

The constraints affecting the economic design of power generation and transmission plant are very different. Thus the use of a transmission system inflexibly tied to the requirements imposed by the generators will, in general, produce less economical power systems.

A.C. transmission over long distances, especially via underground cable, requires frequent shunt compensation and causes stability problems. A.C. interconnections increase the fault level of the overall system. D.C. transmission is free from these problems and has lower losses and design costs.

These advantages were early realised and the idea of generating a.c. power, converting it into d.c. for transmission and converting it back into a.c., was taken seriously. However, the use of static a.c.–d.c. and d.c.–a.c. power conversion is expensive and in general the comparison of alternatives is not straightforward. This subject is discussed in Chapter 9.

Among the many steps which can be identified with the development of the modern h.v.d.c. transmission technology the following are worth mentioning:

(a) A first attempt to combine the advantages of h.v.a.c. turbo-generation and h.v.d.c. transmission was made in the 'twenties' by Calverley and Highfield with the 'transverter'. The idea was based on a number of transformers commutated by brushgear rotating synchronously at high speed.

(b Hewitt's mercury-vapour rectifier, which appeared in 1901, and the introduction of grid control in 1928, provided the basis for controlled rectification and inversion.

(c) Prior to 1940, experiments were carried out in America with thyratrons and in Europe with mercury pool devices.

(d) Countries with long transmission distances like America, the Soviet Union and Sweden, showed great interest in h.v.d.c. developments. In the Soviet Union an experimental single anode valve was constructed during the Second

World War and intensive research was carried out in Sweden from 1940 by the Allmanna Svenska Elektriska Aktiebologet (ASEA).

(e) In the case of Germany, the Secretariat for Aviation encouraged the development of h.v.d.c. technology during the War believing that underground transmission was less vulnerable to air raids.

An experimental transmission system of 15 MW at 100 kV was built between the Charlotenburg and Moabit districts of Berlin. This was intended to be a prototype for a 60 MW, 400 kV system of about 110 km, part of which was built by the end of the War.

(f) After 1945, the interrupted work on valves in the USSR was resumed as part of a wider programme of h.v.d.c. developments. In 1950[2] an underground system was brought into operation between Moscow and Kashira, transmitting 30 MW at 200 kV over a distance of 112 km.

(g) However, it is Sweden that should be credited with having pioneered the development of the modern h.v.d.c. transmission technology as explained in the next section.

1.3 The mercury-arc valve

Although the economic advantage of d.c. power transmission was understood from the early days of the electrical technology, its practical application had to wait for the development of a suitably rated electronic valve.

Among the various switching principles used in the early days of the power electronic industry, mercury-arc rectification was found the most suitable for handling large currents. Multi-phase mercury-pool cathode valves provided with a control electrode (or grid) have been extensively used for over 50 years in industrial and railway applications.

The most successful development of the mercury valve for high voltage applications was carried out in Sweden, where by 1939 Dr. Uno Lamm from ASEA had invented a system of grading electrodes with a single-phase valve construction, which provided the basis for larger peak inverse withstand voltages.

The basic components and main auxiliaries of the ASEA high voltage valve assembly are illustrated in Fig. 1.1.[3,4] In common with earlier technology, this valve included a mercury-pool cathode with a cathode spot, maintained by means of an auxiliary arc, which causes a continuous emission of electrons. It was the anode design that differed greatly from the earlier medium-voltage valve, its main aim being the elimination of the reverse emission of electrons which causes reversal of conduction or arc-back.

With the graded electrodes it was possible to achieve a more uniform distribution of the reverse voltage in the vicinity of the anode. This reduced the energy of the charge carriers striking the anode material and with it the likelihood of arc-backs. The grading electrodes are connected to an external capacitive resistive voltage divider, which together with the interelectrode

4 Development of a h.v.d.c. technology

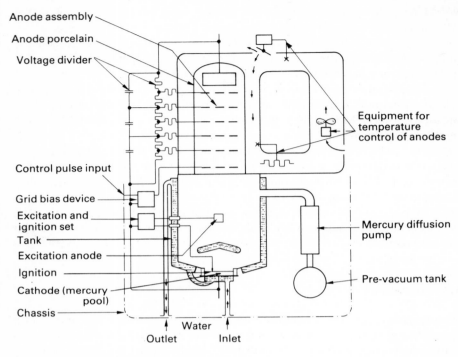

Fig. 1.1 *High voltage mercury-arc valve assembly*

capacitances limits the voltage difference between them to about 5 kV. The anode porcelain forms a vacuum-tight envelope that functions as supporting insulator for the different electrodes in the anode assembly. Depending on the rated current up to six parallel anodes are placed on top of the stainless steel tank.

The quality of the porcelain used for the external cylinder is essential to the viability of the h.v.d.c. valve. Under the influence of the direct voltage component across the valve, some ion migration occurs which causes ion depletion at one end and ion increase at the other end. This effect produces conductivity variation and thus causes uneven voltage distribution. In later designs the use of very high resistance porcelain has reduced this problem dramatically.

Another important problem was the deposit of material throughout the valve which results from charge carriers striking the walls during firings and blockings. This effect appears to limit the maximum direct voltage achieved with mercury-arc bridges to about 150 kV and requires considerable maintenance.

1.4 Mercury-arc schemes

The fast development of the graded electrode mercury-arc valve in the late 1940's paved the way for a first commercial application of a h.v.d.c. technology in 1954. This scheme was immediately followed by others of increased ratings.

The success of the mercury-arc based h.v.d.c. technology can be best illustrated with reference to some of the early schemes. A summary of the main factors justifying their existence is now given and the technical background required to understand these factors will be the subject of subsequent chapters.

1.4.1 Sweden–Gotland link (1954)
A 20 MW, 100 kV d.c. single conduction submarine link to supply power to the island of Gotland was justified economically as an alternative to the establishment of extra thermal generation on the island. The distance (96 km) was too large for transmission by a.c. cable.

1.4.2 English Channel (1961)
A 160 MW, ± 100 kV d.c. submarine link interconnecting Britain and France to take advantage of peak load and generation diversity was marginally justifiable in terms of distance (only 64 km), particularly as the North Sea cannot be used as a return path because of interference with ships' magnetic compasses. However, the relatively small power tie would have been difficult to control with an a.c. cable interconnection, as the British power system has no automatic load-frequency control. The provision of such control throughout the British grid would have been far more expensive than the alternative d.c. scheme.

1.4.3 Volgograd–Donbass (1962–65)
This was the first overhead transmission scheme and was built as a reinforcement of an existing a.c. weak transmission system. The transmission distance is 470 km and the link rating 720 MW at ± 400 kV and 900 A. The scheme used an alternative design of mercury-arc valves with a single (air-cooled) anode and an oil-cooled tank. The bi-polar scheme uses four bridges per pole and two valves in series each for 50 kV and 900 A.

1.4.4 New Zealand link (1965)
A 600 MW, ± 250 kV d.c. partly overhead (570 km) and partly submarine (40 km) link was installed to transfer to the North Island the surplus of energy from the hydro resources in the South Island. Given the relatively short submarine interconnection the scheme was feasible using conventional a.c. technology. However the depth of the Cook Strait would have restricted the

use of a.c. cable to relatively low voltages and thus resulted in many parallel circuits. Under such conditions and given the reasonably large overhead transmission the d.c. solution was found more economical.

1.4.5 Konti–Skan (1965)

This is a monopolar 250 MW, 250 kV d.c. (using earth return), partly overhead, partly submarine interconnection between Sweden and Denmark. The cable section, although sufficiently long (87 km), is divided in two parts and an a.c. interconnection would have been feasible. In fact the economic justification of d.c. was very marginal, but it offered the advantage of further future extension (by means of a second pole) to 500 MW, a power rating beyond the stability limit of an alternative a.c. interconnection between the two countries.

1.4.6 Sakuma interconnection (1965)

This is a zero-distance frequency interconnector to interchange up to 300 MW (at $\pm$ 125 kV) in either direction between the 50 and 60 Hz power systems in Japan in case of a.c. network disturbances or lack of energy in either of the systems. In this case there was no practical alternative to the a.c.–d.c.–a.c. converting station.

1.4.7. Sardinia–Italy (mainland) (1967)

The monopolar 200 MW at 200 kV d.c. (with earth and sea return) was mainly installed to provide frequency support to the Sardinian a.c. network by power/frequency control of the d.c. link.

Transmission alternates between submarine cable and overhead line as it passes through Sardinia, the sea, Corsica, the sea and the Italian mainland. The total cable distance of 121 km justified the use of d.c.

1.4.8 Pacific intertie (1970)

Unlike the other schemes discussed above, the Pacific d.c. intertie runs in parallel with two other a.c. circuits at 500 kV (60 Hz).

This d.c. scheme was partly justified in terms of distance (1372 km of overhead transmission) and partly by the availability of additional control features which were expected to help damp out the power oscillations experienced in the parallel a.c. transmission system already in existence. It consisted of a 1440 MW at $\pm$ 400 kV bidirectional link to take advantage of the seasonal diversity in load and generation between the northwest and southwest areas of the United States.

1.4.9 Kingsnorth scheme (1974)

The main purpose of the 640 MW (at $\pm$ 266 kV) three-terminal Kingsnorth–Beddington–Willesden d.c. scheme in the U.K. was the reinforcement of an existing a.c. system in areas of high load density without increasing the short-circuit level. It would have been difficult to justify the use of d.c. purely in terms of distance, as the two underground transmission distances were very marginal (59 and 82 km respectively).

1.4.10 Nelson River Bipole 1 (1973–77)
Perhaps this was the more obvious application of h.v.d.c. transmission as it involves the transport of bulk power from distant hydro generation to the loading centre (in this case Winnipeg). The distance involved is 895 km with a nominal capacity of 1620 MW (at ± 450 kV) to be expanded to its ultimate capacity of about 6500 MW. The first stage of the Nelson River scheme used the highest power mercury-arc valves ever developed (i.e. 150 kV, 1800 A); this stage also made history by being the last case of successful competition for the mercury-arc over the solid state technologies.

1.5 The solid state technology

In parallel with the amazing development of the micro-electronic technology of recent times there has been an equally impressive, though much less publicised, macro-electronic revolution in the power field sharing the same basic ingredients, i.e. switching and silicon. The main exponent of the macro-electronic technology must surely be the solid state h.v.d.c. convertor which will be described in some detail in Chapter 6.

The appearance of the thyristor, or silicon controlled rectifier (SCR), in the late 1950's had a dramatic effect on the static convertor technology. The thyristor (Fig. 1.2) is a silicon four-layer (p.n.p.n.) three-junction device which performs as a controllable rectifier. The complete voltage–current characteristic of a thyristor unit is illustrated in Fig. 1.3. A small current injection through the gate terminal makes a forward-biased thyristor switch from a very high to a very low impedance state, thus approximating the characteristic of the ideal switch, with practically unlimited amplification factor.

Two-terminal breakover (in the absence of gate injection) can also take place, either by sufficient forward (anode–cathode) bias (V_{BO}) or excessive rate of change of voltage dV/dt. Once the thyristor is turned on, it can only be turned off, or blocked, by reducing the main circuit current below a very low critical value, called the holding current. While gate turn-off thyristors are feasible, they have not been proved economical for high power applications so far.

Under reverse bias there is a critical breakdown level (V_{BD}), below which the thyristor behaves like a p.n. junction diode, i.e. with only a low leakage current.

The thyristor can be destroyed by excessive reverse voltage (V_{BD}), extended overcurrents and excessive rate of change of current (di/dt). Therefore, when connected in series, the individual devices have to be properly protected against overvoltages, overcurrents, di/dt and dv/dt. This subject will be further discussed in later chapters.

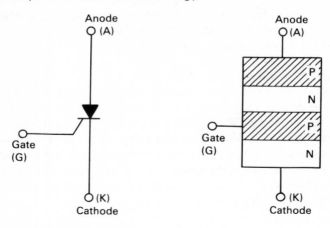

Fig. 1.2 *Basic structure and symbol of a thyristor*

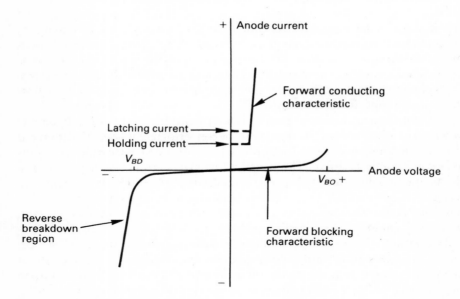

Fig. 1.3 *Thyristor V–I static characteristic*

For the economical use of thyristors in h.v.d.c. transmission, it was necessary to improve the critical breakdown level (i.e. the blocking-voltage rating), and blocking-voltages of 5000 V are now in commercial production. Higher voltages are feasible, but normally at the expense of severe de-rating on other important parameters. The main limitation of the higher voltage thyristor is a greater requirement of forward recovery time, which results in larger nominal extinction angles (γ) during invertor operation; some thyristor schemes use larger steady state γ's than the latest mercury-arc schemes.

From a few kilowatts in its origin, the thyristor quickly grew into the megawatt range and expanded from the power utilisation to the power transmission fields at an unprecedented rate.

Although the ratings of the individual devices are still increasing, there is no need for dramatic growth in this respect because of the progress made in thyristor architecture; the individual devices can now be connected in series, parallel and multibridge configurations to process practically unlimited power.

Thyristor valves can now be designed to perform all the switching, control and conversion duties of the mercury-arc valve and generally much more economically.

1.6 Thyristor take-over

In spite of the successful operation of the mecury-arc schemes, the incidence of arc-backs, considerable maintenance and voltage limitations encouraged the development of the solid state technology.

The impact of the thyristor technology was not just due to valve cost (which initially was much the same as the cost of the mercury-arc valve) but also to the overall economic effect of station plant and layout.

Even before the commissioning of the last two mercury-arc schemes (Nelson River and Kingsnorth) the small experience gained with thyristor valves was sufficient to discourage any further development of the mercury-arc technology.

The technical reasons in favour of solid state h.v.d.c. transmission are basically the same as those given for the mercury-arc schemes discussed above. However, a brief look at some of the solid state schemes will help to understand the incredible progress already made with the new technology.

1.6.1 Eel River (1972)

If the Nelson River made history for being the last of the old technology, another Canadian River (the Eel) started the new technology. To be fair, solid state additions to previously existing schemes had already been operating in Sweden and the U.K. However, the Eel River scheme (350 MW at 80 kV) was the first large h.v.d.c. project specifically designed for thyristor valves. It is a back to back asynchronous interconnection of two systems (New Brunswick and Hydro-Quebec) of nominally equal frequency, but drifting in relation to each other.

Each station consists of two convertor bridges with a total of 4800 air cooled thyristors placed in changeable 40-unit modules. It was necessary to parallel four thyristor units per convertor arm.

1.6.2 Cabora–Bassa (1977–79)

A decision was taken in 1969 to use solid state valves to transmit 1920 MW at

± 533 kV between the Zambia River in Mozambique and Johannesburg, separated by 1414 km. This was the first scheme, whether a.c. or d.c., in the megavolt range (between poles). It was also the first case of international bulk power transmission. The first stage of the scheme started commercial operation in 1977. It uses over 36 000 thyristors, oil-colled and oil-insulated in an outdoor valve layout which involves four bridges per pole at each end of the link. It is interesting to consider that switching, which used to be the curse of power transmission circuits, had become the basis of power controllability. In fact, the normal operation of the complete Cabora–Bassa scheme will need of the order of 4×10^6 ON and OFF individual switchings per second.

1.6.3 Inga–Shaba (1981)

The fast progress made in thyristor architecture became apparent in this project, approved by the Zaire Government in 1974 to exploit the Zaire River hydro potential. In the first stage the d.c. system transmits 560 MW over 1700 km at ± 500 kV. Additional convertors will be connected in parallel on the d.c. side in stages 2 and 3 increasing the rated current from 560 to 1120 A and 2240 A respectively. Thus for the first time in an h.v.d.c. project the Inga–Shaba terminal stations will have convertors operating in parallel. The scheme includes double-valves of a modular design and air-insulated for indoor installation.

1.6.4 Skagerrak (1976–77)

A further step from the Inga–Shaba project led to this scheme where the four individual valves (per phase) of a twelve-pulse convertor are combined into a single vertical quadruple valve, also of modular design for easy maintenance. This arrangement has become standard in later schemes.

The Skagerrak scheme provides 500 MW (1000 in the second state) of interconnection between Norway and Denmark. It includes 127 km of cable transmission and 113 km of overhead line. D.C. was justified partly in terms of distance and partly by the different capacities of the Norwegian and European systems.

1.6.5 Square Butte (1977)

The cost of a generating plant in North Dakota and electrical energy transport by a 750 km h.v.d.c. link between North Dakota and Minnesota was found a cheaper alternative to the cost of coal transportation for use of local generation in Minnesota. Moreover, the use of d.c. offered a solution to the system stabilization problem of an alternative a.c. transmission scheme.

The thyristor valves of the Eel River scheme had been shown to possess considerably higher availability than the rest of the circuitry needed for 6-pulse operation. This led to the selection of a 12-pulse convertor per pole with a rating of 500 MW ± 250 kV.

1.6.6 Nelson River Bipole 2 (1978–85)

The final stage of the Nelson River Bipole 2 scheme is rated 1800 MW at ± 500 kV. The 12-pulse group quadruple valves in this scheme are water cooled. With the experience of this scheme water is now generally regarded as the best coolant and it is likely to replace air cooling from now on in large rated convertors.

1.6.7 Itaipu scheme

The culmination of the solid state technology is the Itaipu h.v.d.c. scheme which involves some 6000 MW of d.c. power transmission between Paraguay and Brazil over a distance of 850 km. The total eventual transmission is 12 600 MW, half of which will use 800 kV a.c. The compromise reached was based on the different frequencies of the two countries (50 and 60 Hz) and the agreement for each country to have the right to use half of the power generated at Itaipu. Consequently half of the generators are designed for 50 Hz and half for 60 Hz. The scheme, still under construction at the time of writing, uses two bipolar d.c. lines at ± 600 kV, each consisting of four (two per pole) twelve pulse water-cooled valve convertors and the power rating of the individual valves is 788 MW.

1.6.8 Ekibastuz–Centre

At present under construction, this intercontinental scheme will eventually carry 6000 MW of power generated at the mine-mouth thermal power stations being erected in the Ekibastuz coal field region in Siberia to the European part of the Soviet Union. The main features of the scheme are the transmission distance (2414 km) and the very high transmission voltage (1500 kV at ± 750 kV). The scheme is expected to be commissioned before 1985.

1.6.9 New cross-channel link

Almost 20 years after the commissioning of the first cross-channel link between England and France, an agreement to proceed with the construction of a 2000 MW link was issued by EDF and CEGB in 1981. The convertor stations will be linked by eight cables operating at ± 270 kV. The scheme will consist of two 1000 MW bipoles between Sellindge in S.E. England and Bonningues-les-Calais in Northern France; the first bipole is planned for completion in 1985 with the second in 1986.

The main novel feature of the scheme is the use of different convertor equipment (with valves and control of different manufacture) at each end of the link. On the English side the convertor plant will include three high-speed static compensators of the saturated reactor type to provide, in conjunction with switched capacitor banks, the required reactive power control capability for load and load rejection conditions.

12   Development of a h.v.d.c. technology

1.7 Operation reliability

A recent report on the operational performance of h.v.d.c. schemes summarises the experience with h.v.d.c. plant unavailability as follows[5]:

(*a*) The mercury-arc valves performance depends on their internal condition and external auxiliary equipment, and their availability has been good whenever a high level of maintenance has been provided which included the provision of spare valves and skilled staff.
(*b*) The influence of the convertor valves (whether of the mercury-arc or thyristor type) on forced unavailabilities has been very small. Most of the lengthy outages in h.v.d.c. schemes have been due to failure of other large plant components such as the convertor transformers, a.c. filter circuits and synchronous compensators.
(*c*) Excluding transmission line outages, the forced unavailability of the thyristor convertor plant has been only about a third of that of mercury-arc schemes.

The relatively small experience with thyristor schemes justifies the present conviction that the thyristor valve is a more reliable device. Typical thyristor failure rates of 0·6 per cent per operating year are currently achieved but even such failures do not affect thyristor valve operation, due to built-in thyristor redundancy; i.e. the faulty thyristors (or auxiliary equipment) can wait for replacement until the next scheduled maintenance period.

1.8 Future expansion

A clear indication of the expansion of the mercury-arc and thyristor h.v.d.c. technologies is given in Fig. 1.4. At the beginning of 1980, after 25 years of commercial existence, the total installed capacity of h.v.d.c. schemes was 12 000 MW, divided equally between mercury-arc and thyristor based schemes. The thyristor had matched the mercury-arc capacity in only eight years of existence.

With a further 20 000 MW capacity of thyristor schemes expected to be installed by the mid-eighties, the expansion of the h.v.d.c. technology points exclusively in this direction.

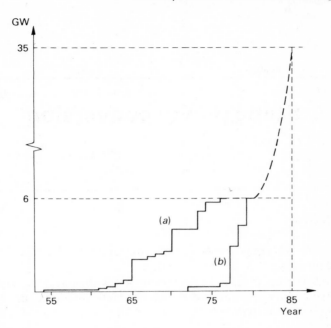

Fig. 1.4 *Installed and projected capacity of h.v.d.c. schemes*
(a) Mercury-arc schemes
(b) Thyristor schemes

1.9 References

1 The history of high voltage direct current power transmission', *Direct Current*: Part I, December 1961, p. 260; Part II, March 1962, p. 60; Part III, September 1962, p. 228; Part IV, January 1963, p. 2; Part V, April 1963, p. 89.
2 PIMENOV, V. P. (1957): 'The work of the Direct Current Institute (Leningrad)', *Direct Current*, Vol. 3, No. 6, pp. 185–191.
3 LAMM, U. (1964): 'Mercury-arc valves for high voltage d.c. transmission', *Proc. IEE*, Vol. III, No. 10, pp. 1747–1753.
4 BERNERYD, S. and FUNKE, B. (1966): 'Design of high voltage mercury-arc valves', *IEE Conference on High Voltage D.C. Transmission*, Publication 22, Manchester.
5 RUMPF, E. (1980): 'The operational performance of h.v.d.c. systems throughout the world during 1975–1978', *1980 Symposium sponsored by the Division of Electric Energy Systems USDOE*, Phoenix, Arizona, pp. 1–23.

Static power conversion
Chapter 2

2.1 Introduction

The static conversion of power from a.c. to d.c. and from d.c. to a.c. constitutes the central process of h.v.d.c. transmission.

It is therefore important to begin the subject with a clear understanding of the conversion principles, and of the steady state relationships, which exist between the various parameters involved in the process of static power conversion.

This chapter describes the requirements of stable convertor operation, the effect of controlled rectification and the commutation phenomena. Detailed consideration is given to the voltage and current waveforms, and to the reactive power demand and harmonic problems attached to convertor operation.

2.2 Basic conversion principle

The first consideration to be made in the process of static power conversion is how to achieve instantaneous matching of the a.c. and d.c. voltage levels, given the limited number of phases and switching devices which are economically feasible.[1] With reference to Fig. 2.1(a), in the absence of energy storing elements on either side and in the presence of a constant d.c. voltage, the time variation of the a.c. voltage waveform and any voltage supply deviations from the nominal level, will cause theoretically infinite current level transients.

For practical operation, enough series impedance must therefore be included to absorb the continuous voltage mismatch between the two sides. If such an impedance is exclusively located on the a.c. side (Fig. 2.1(b)), the switching devices transfer the instantaneous direct voltage to the a.c. system according to transformer connection and ratio; thus the circuit configuration is basically a voltage convertor, with the possibility of altering the d.c. current by thyristor control.

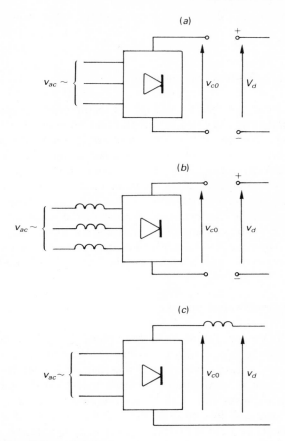

Fig. 2.1 *A.C./D.C. voltage matching*
(a) Unmatched circuit
(b) Circuit for voltage conversion
(c) Circuit for current conversion

If a large smoothing reactor is placed on the d.c. side (Fig. 2.1(c)), only pulses of constant direct current flow through the switching devices into the transformer secondary windings. These current pulses are then transferred to the primary side according to transformer connection and ratio; thus a basically current convertor results, with the possibility of adjusting the direct voltage by thyristor control.

The use of voltage conversion was rejected in mercury-arc convertors due to the impossibility of recovering from arc-back disturbances. Even with thyristor schemes, rapid changes in the supply voltage can only be accommodated within narrow limits and require the use of large series impedances, which would be uneconomical in terms of reactive power compensation. Therefore the current

16   Static power conversion

conversion principle is generally accepted as the basis of h.v.d.c. convertor design.

2.3 Selection of convertor configuration

The three-phase bridge, shown in Fig. 2.2, is the only configuration used in h.v.d.c. transmission. As compared with other possible alternatives, such as the three-phase double star or the six-phase diametrical connections, the bridge configuration provides better utilisation of the convertor transformer and a lower peak inverse voltage across the convertor valves.[2]

As the figure shows, two valves are connected to each phase terminal, one with the anode connected to it (shown on the upper side of the bridge) and the other with the cathode connected to it (shown on the lower side of the bridge). The need for two valves conducting in series is not a drawback in high voltage applications, particularly with solid state convertors, because of the need for many series connected units to withstand the voltage levels used.

2.4 The ideal commutation process[2]

(To understand the operation of a three-phase bridge rectifier let us first consider the idealised case of a convertor bridge connected to an infinitely strong power system (i.e. of zero source impedance). Under this condition, the transfer of current (commutation) between valves on the same side of the bridge takes place instantaneously.)(The switching sequence and the rectified voltage waveform are illustrated in Fig. 2.2 for the case of an uncontrolled bridge rectifier (i.e. on diode operation); valves 1, 3, 5 at the top and 4, 6, 2 at the bottom are connected to phases red, yellow and blue respectively.)

With reference to Fig. 2.2(a) and (g), and starting at instant A, phases R and Y are involved through conducting valves 1 and 6. This operating state continues up to point B, after which valve 2 becomes forward-biased, since its anode, directly connected to that of valve 6, is positive with respect to its cathode (connected to phase blue); therefore at point B the current commutates naturally from valve 6 to valve 2 (Fig. 2.2(b)).

A similar argument applies at point C, with reference to valves 1 and 3 on the upper half of the bridge. The anode of valve 3 (connected to Y phase) begins to be positive with respect to its cathode (connected to phase R through the conducting valve 1) and a commutation takes place from valve 1 to valve 3 (Fig. 2.2(c)). This is followed by commutation from valve 2 to valve 4 at point D, valve 3 to valve 5 at point E, valve 4 to valve 6 at point F, and valve 5 to valve 1 at point G. This completes the switching cycle sequence.

The output waveform in Fig. 2.2(g) shows the voltage variation of the positive (common cathode) and negative (common anode) poles with respect

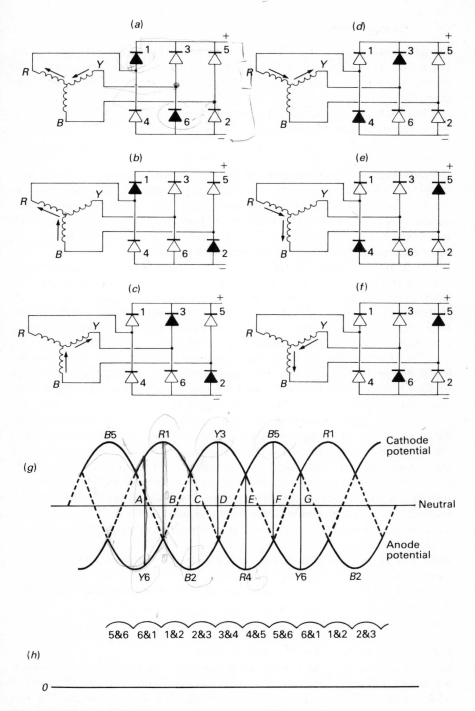

Fig. 2.2 Bridge conducting sequence and d.c. voltage waveforms

to the transformer neutral. Fig. 2.2 (*h*) shows the output voltage, i.e. the voltage of the positive pole with respect to the negative pole. It is seen that the output voltage has a ripple, or harmonic frequency, of six times the main frequency.

Each valve carries the full value of direct current for one third of the cycle, and there are always two valves conducting in series.

2.4.1 Effect of gate control

By delaying the firing instants of the valves with respect to the voltage crossings, the commencement of the natural commutations described in Section 2.4 can be delayed by a definite time interval and the effect of this action on the direct voltage waveforms is illustrated in Figs. 2.3(*a*) and (*b*). It is noticeable that the voltage area, and therefore the mean direct voltage, are reduced in proportion with the magnitude of the delay.

For delay angles above 60° some negative voltage periods begin to appear. If the bridge output were connected to a pure resistance, the bridge unidirectional current conduction property would prevent reverse current flow during these negative voltage periods, and the operation would then be intermittent. However, the provision of a large smoothing reactor maintains positive current flow during the negative periods, and energy is transferred from the reactor magnetic field to the a.c. system. The voltage waveforms for a delay of 90°, illustrated in Figs. 2.4(*a*) and (*b*), show equal positive and negative voltage regions (indicated by horizontal and vertical shaded areas respectively); the mean direct voltage is therefore zero with a 90° delay.

Beyond 90° the mean voltage is negative and the bridge operation can only be maintained in the presence of a d.c. power supply. This supply overcomes the negative voltage and forces the current to conduct in the same direction (i.e. from anode to cathode), in opposition to the induced e.m.f. in the convertor transformer. This indicates that power is being supplied to the a.c. system, i.e. the convertor is inverting. Figs. 2.5(*a*) and (*b*) illustrate the (ideal) limiting case of full inversion which would require a delay of 180°.

Three conditions are thus required to permit power inversion:

(*a*) an active a.c. voltage source which provides the commutating voltage waveforms;
(*b*) provision of firing angle control to delay the commutations beyond $\alpha = 90°$;
(*c*) a d.c. power supply.

2.4.2 Valve current and voltage waveforms

In the presence of a large smoothing reactor on the d.c. side, the voltage waveform of Fig. 2.3(*b*) will produce a constant direct current, the level of which depends on the mean voltages at both ends of the link and the link resistance. For the idealized commutating conditions described in Section 2.4,

Static power conversion 19

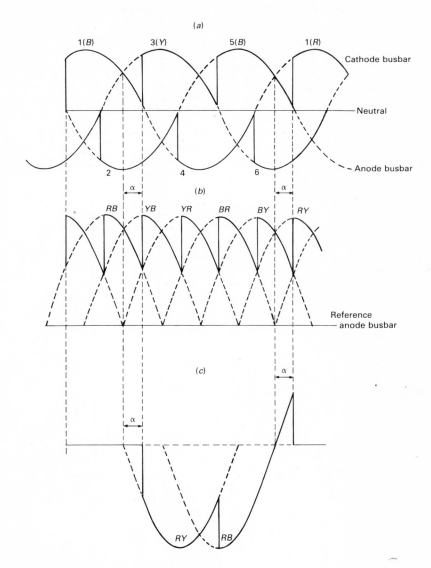

Fig. 2.3 Effect of firing delay on voltage waveforms
(a) Common anode and common cathode voltages
(b) Direct voltage
(c) Voltage across valve 1

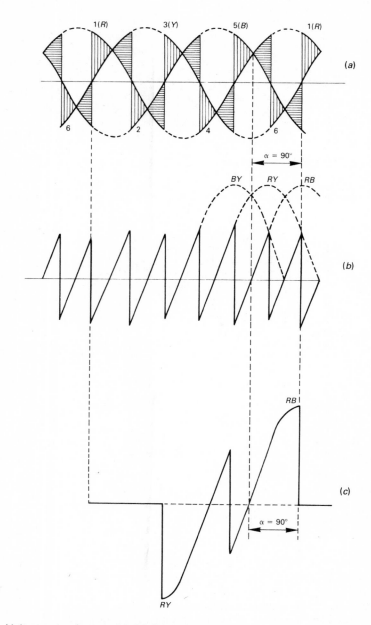

Fig. 2.4 *Voltage waveforms with 90° firing delay*
(a) Common anode and common cathode voltages
(b) Direct voltage
(c) Voltage across valve 1

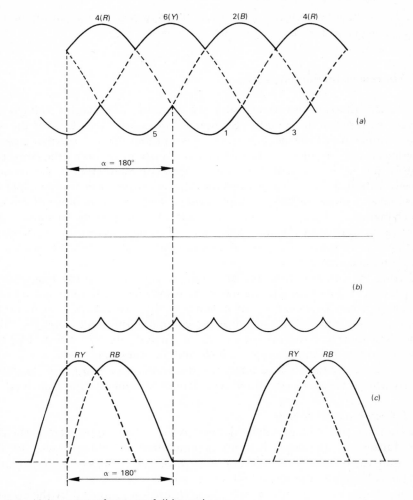

Fig. 2.5 *Voltage waveforms on full inversion*
(a) Common anode and common cathode voltages
(b) Direct voltage
(c) Voltage across valve 1

the valve current will be a rectangular pulse lasting 120°, its relative position with reference to the corresponding voltage waveform being determined by the firing delay angle α.

During the conducting period, and disregarding the small internal voltage drop, the voltage across the valve is zero. Outside the conducting period, the voltage across the valve consists of varying portions of two phase-to-phase voltage waveforms; these are the phase to which the valve is directly connected and the phase of the conducting valve on the same side of the bridge. Examples of valve voltage waveforms for different delay angles are shown in Fig. 2.3(c), 2.4(c) and 2.5(c).

2.5 The real commutation process

In practice, the zero impedance supply required to produce the voltage and current waveforms described in Section 2.4 does not exist. Even if the a.c. system impedance were negligible, there is considerable transformer leakage reactance between the convertor and the a.c. system. In theory the convertor transformer is not essential to the process of static power conversion. However there are practical reasons for using convertor transformers, like the possibility of phase-shifting multiple bridges and the availability of on-load tap-changing, which will become apparent when discussing harmonics and reactive power compensation later on. The main effect of a.c. system reactance is to reduce the rate of change of current or, in other words, to lengthen the commutating time.

During the commutation, the magnetic energy stored in the reactance of the previously conducting phase has to be transferred to the reactance of the incoming phase. That energy only depends on the direct current level and the inductance per phase. The speed of the commutation process, on the other hand, and therefore the rate of change of current, are also affected by two other parameters, i.e. the supply voltage and the firing delay angle.

It is thus essential, before analysing the commutation process, to define clearly what is meant by commutating voltage and commutation reactance.

2.5.1 Commutating voltage

The commutating voltage can be defined as the voltage appearing on the d.c. line during the periods when no commutation is taking place. In this operating region only direct current flows through the a.c. system impedance, and the voltage waveform is therefore sinusoidal.

Considering the a.c. current waveform distortion produced by the convertor, it will be necessary to go back to the system source voltage to find an undistorted supply to the convertor. In practice, however, phase-shifting and filtering are provided with every h.v.d.c. convertor station and the voltage waveform at the filter busbar is reasonably sinusoidal (under steady state and normal operating conditions!). Such voltage can therefore be used as the commutating voltage.

2.5.2 Commutation reactance[3]

The commutation reactance can be defined as the reactance between the commutating voltage, as defined in Section 2.5.1, and the convertor valves. Figure 2.6 shows the general case of n bridges connected in parallel on the a.c. side. In the absence of filters, pure sinusoidal voltages only exist behind

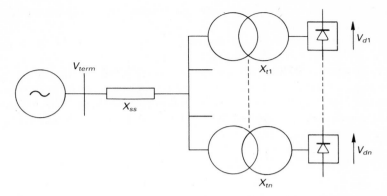

Fig. 2.6 'n' bridges connected in series on the d.c. side and in parallel on the a.c. side

the system source impedance (X_{ss}) and in such case the commutation reactance (X_{cj}) for the jth bridge is given by

$$X_{cj} = X_{ss} + X_{tj}. \tag{2.1}$$

If the bridges are under the same controller, or under identical controllers, it is preferable to create a single equivalent bridge. The commutation reactance of such equivalent bridge depends upon the d.c. connections and on the phase-shift between the bridges.

Bridges with the same phase-shift will commutate simultaneously and the equivalent reactance must reflect this. For a series connection of k bridges the commutation reactance of the equivalent bridge is:

$$X_c(\text{series}) = kX_{ss} + X_{tj}, \tag{2.2}$$

where j represents any of the n bridges.

If the bridges are connected in parallel on the d.c. side the equivalent bridge commutation reactance is:

$$X_c(\text{parallel}) = X_{ss} + \frac{1}{k}X_{tj}. \tag{2.3}$$

With perfect filtering, or with a combination of filters and transformer phase-shift (refer to Section 2.5.1), the voltage on the a.c. side of the convertor transformers may be assumed to be sinusoidal and hence X_{ss} has no influence on the commutation. It must be noted that h.v.d.c. schemes are normally designed for 12-pulse operation and that filters are always provided (i.e. the system impedance can be ignored).

However, the presence of local plant components at the convertor terminals, such as synchronous compensators, will affect the commutation reactance. Since the commutation produces a phase-to-phase short circuit during a small fraction of a cycle, the synchronous machines must be represented by their subtransient reactances.

By way of example let us consider the two ends of the New Zealand h.v.d.c. link (with reference to Fig. 2.7 and 2.8).

(a) At the receiving (Haywards) end (Fig. 2.7) the effect of the subtransient reactance of the synchronous compensators on the tertiaries of the convertor transformers must be taken into account. The approximate equivalent circuit is illustrated in Fig. 2.7(b) and the commutation reactance is

$$X_c = X_s + \frac{X_p(X_t + X_d'')}{X_p + X_t + X_d'} \tag{2.4}$$

where X_s is the transformer secondary leakage reactance, X_p is the transformer primary leakage reactance, X_t is the transformer tertiary leakage reactance, and X_d'' is the subtransient reactance (direct axis) of the synchronous condenser unit.

(b) At the sending (Benmore) end (Fig. 2.8) the subtransient reactance of the generators is combined in parallel with the secondary reactance of the interconnecting transformer. The primary reactance is beyond the filters and can thus be neglected. An approximate equivalent circuit is illustrated in Fig. 2.8(b). Although there are two convertor groups commutating on this reactance, the commutations are not simultaneous due to the 30° phase-shift of their respective transformers. Thus the effective commutation reactance per group is

$$X_c = X + \frac{X_d'' X_s}{X_d'' + nX_s}, \tag{2.5}$$

where X is the two-winding transformer leakage reactance, X_s is the interconnecting transformer secondary leakage reactance (note filters connected to tertiary winding), X_d'' is the generator subtransient reactance, and n is the number of generators connected. Equation (2.5) is only valid for commutation angles not exceeding 30°; however this covers most operating conditions.

2.5.3 Analysis of the commutation circuit

Let us consider the commutation process between valves 1 and 3 of a convertor bridge, connected to a system with a source voltage v_c, a commutation reactance per phase X_c and negligible source resistance.

With reference to Fig. 2.9, the commutation from valve 1 to valve 3 can start (by the firing of 3) any time after the upper voltage crossing between v_{CR} and v_{CY} (and must be completed before the lower crossing of these two voltages).

Static power conversion 25

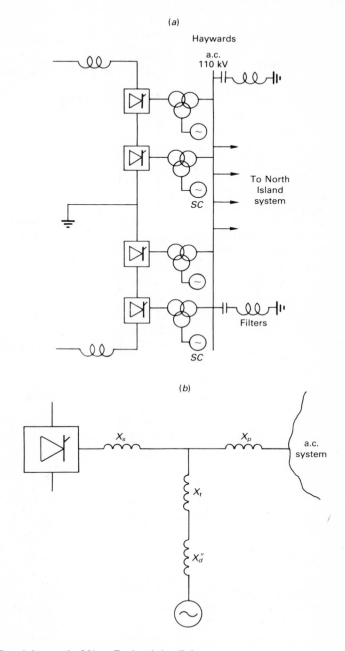

Fig. 2.7 *Receiving end of New Zealand d.c. link*
(a) Simplified diagram
(b) Equivalent circuit to calculate the commutation reactance

26   Static power conversion

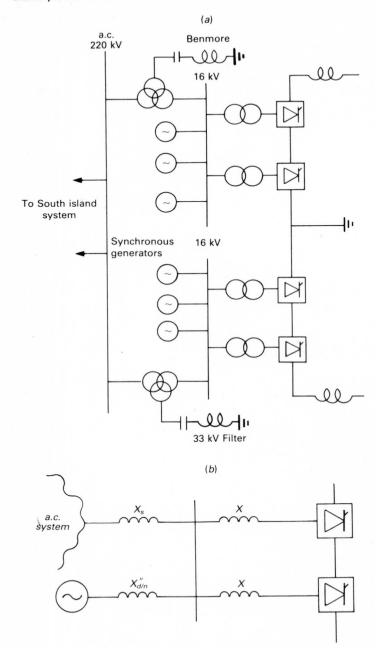

Fig. 2.8 *Sending end of the New Zealand d.c. link*
  (a) Simplified diagram
  (b) Equivalent circuit to calculate the commutation reactance

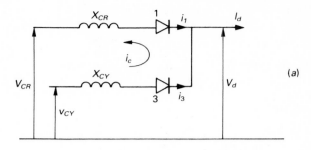

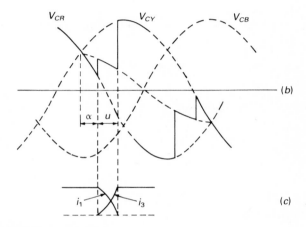

Fig. 2.9 *The commutation process*
(a) Equivalent circuit of the commutation from valve 1 to valve 3
(b) Voltage waveforms showing an early (rectification) and late (inversion) commutations
(c) The commutating currents

Since $v_{CY} > v_{CR}$, a commutating current $i_c\,(=i_3)$ builds up at the expense of i_1 so that at all times

$$i_1 + i_3 = I_d.$$

As the rates of change of i_3 and $-i_1$ are equal (provided that the commutation reactances are balanced), the voltage drops across X_{CR} and X_{CY} are the same and thus, during the overlap period, the direct voltage v_d is the mean value of v_{CY} and v_{CR}.

From the circuit of Fig. 2.9(a) and assuming $X_{CR} = X_{CY} = X_C$ we can write

$$v_{CY} - v_{CR} = 2(X_c/\omega)\,d(i_c)/dt. \tag{2.6}$$

Taking as a reference the voltage crossing between phases R and Y

$$v_{CY} - v_{CR} = \sqrt{2}V_c \sin \omega t,$$

where V_c is the phase-to-phase rms voltage.

Equation (2.6) can also be written as

$$\frac{1}{\sqrt{2}} V_c \sin(\omega t)\, d(\omega t) = X_c\, di_c. \tag{2.7}$$

and integrating from $\omega t = \alpha$

$$\frac{1}{\sqrt{2}} \int_\alpha^{\omega t} V_c \sin(\omega t)\, d(\omega t) = X_c \int_0^{i_c} d(i_c). \tag{2.8}$$

The instantaneous expression for the commutating current is thus

$$i_c = \frac{V_c}{\sqrt{2} X_c}[\cos\alpha - \cos(\omega t)] \tag{2.9}$$

and substituting the final condition, i.e. $i_c = I_d$ at $\omega t = \alpha + u$ yields

$$I_d = \frac{V_c}{\sqrt{2} X_c}[\cos\alpha - \cos(\alpha + u)]. \tag{2.10}$$

2.6 Rectifier operation

Typical voltage and current waveforms of a bridge operating as a rectifier with the commutation effect included are shown in Fig. 2.10, where P indicates a firing instant (e.g. P_1 is the firing instant of valve 1), S indicates the end of a commutation (e.g. at $S5$ valve 5 stops conducting), and C is a voltage crossing (e.g. $C1$ indicates the positive crossing between phases blue and red).

Figure 2.10(a) illustrates the positive (determined by the conduction of valves 1, 3, 5) and negative (determined by the conduction of valves 2, 4, 6) potentials with respect to the transformer neutral. Figure 2.10(b) shows the direct voltage output waveform.

The potential across valve 1, also shown in Fig. 2.10(b), depends on the conducting valves. When valve 1 completes the commutation to valve 3 (at $S1$) the voltage across will follow the red–yellow potential difference until $P4$. Between $P4$ and $S2$ the commutation from valve 2 to valve 4 (see Fig. 2.10(a)) reduces the negative potential of phase red and causes the first voltage dent. The firing of valve 5 (at $P5$) increases the potential of the common cathode to the average of phases yellow and blue; this causes a second commutation dent, at the end of which (at $S3$) the common cathode follows the potential of phase blue (due to the conduction of valve 5). Finally the

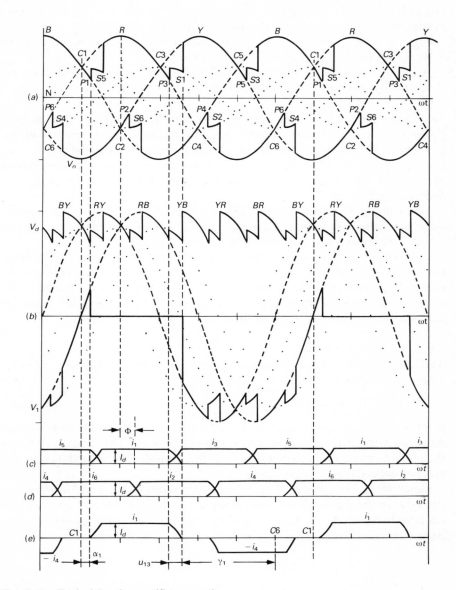

Fig. 2.10 *Typical 6-pulse rectifier operation*
(a) Positive and negative direct voltages with respect to the transformer neutral
(b) Direct bridge voltage V_d, and voltage across valve 1
(c), (d) Valve currents i_1 to i_6
(e) A.C. line current of phase R

commutation from valve 4 to valve 6 (between *P*6 and *S*4) increases the negative potential of valve 1 anode and produces another voltage dent.

Figures 2.10(*c*) and (*d*) illustrate the individual valves (1 and 4) and Fig. 2.10(*e*) the phase (red) currents respectively.

A number of reasonable approximations have to be made to simplify the derivation of the steady state equations that follow. These are:

(*a*) The convertor valves are treated as ideal switches. When calculating the power loss, the valve resistance can be added to that of the d.c. transmission line.
(*b*) The a.c. systems consist of perfectly balanced and sinusoidal e.m.f.'s, the commutation reactances are equal in each phase and their resistive components are ignored. The main effects of non-ideal supply waveforms are discussed in Section 2.9.2.
(*c*) The direct current is constant and ripple-free, i.e. the presence of a very large smoothing reactor is assumed. The effect of non-ideal d.c. current waveforms is discussed in Section 2.9.2.
(*d*) Only two or three valves conduct simultaneously, i.e. two simultaneous commutations are not considered. The low a.c. voltage and/or high d.c. current required to cause simultaneous commutations are prevented in the steady state; during disturbances, on the other hand, the convertor behaviour can only be predicted by dynamic analysis.[8]

2.6.1 Mean direct voltage

The following expression can be easily derived for the average output voltage with reference to the waveforms of Fig. 2.10.

$$V_d = (1/2)V_{c0}[\cos\alpha + \cos(\alpha + u)], \tag{2.11}$$

where V_{c0} is the maximum average d.c. voltage (i.e. at no-load and without firing delay); for the three-phase bridge configuration $V_{c0} = (3\sqrt{2}/\pi)V_c$, and V_c is the phase to phase rms commutating voltage referred to the secondary (or valve side) of the convertor transformer.

Equation (2.11) specifies the d.c. voltage in terms of V_c, α and u. However, the value of the commutation angle is not normally available and a more useful expression for the d.c. voltage, as a function of the d.c. current, can be derived from eqns. (2.10) and (2.11), i.e.

$$V_d = V_{c0}\cos\alpha - \frac{3X_c}{\pi}I_d. \tag{2.12}$$

2.6.2 A.C. current

The rms magnitude of a rectangular current waveform (neglecting the commutation overlap) is often used to define the convertor transformer MVA, i.e.

Static power conversion 31

$$I_{\text{rms}} = \sqrt{\{(1/\pi) \int_{-\pi/3}^{\pi/3} I_d^2 d(\omega t)\}} = \sqrt{2} I_d / \sqrt{3}. \tag{2.13}$$

Since harmonic filters are normally provided at the convertor terminals, the current flowing in the a.c. system contains only fundamental component frequency and its rms magnitude (obtained from the Fourier analysis described in Section 2.9) is

$$I_1 = I_d \sqrt{6}/\pi. \tag{2.14}$$

If the effect of commutation reactance is taken into account, the current waveform for a star/star transformer connection is shown in Fig. 2.10(e). Using eqns. (2.9) and (2.10) the currents of the incoming and outgoing valve during the commutation are defined by eqns. (2.15) and (2.16) respectively:

$$i = \frac{I_d(\cos \alpha - \cos \omega t)}{\cos \alpha - \cos(\alpha + u)} \quad \text{for} \quad \alpha < \omega t < \alpha + u, \tag{2.15}$$

$$i = I_d - I_d \frac{\cos \alpha - \cos(\omega t - 2\pi/3)}{\cos \alpha - \cos(\alpha + u)} \quad \text{for}$$

$$\alpha + \frac{2\pi}{3} < \omega t < \alpha + \frac{2\pi}{3} + u. \tag{2.16}$$

In between commutations the current is

$$i = I_d \quad \text{for} \quad \alpha + u < \omega t < \frac{2\pi}{3} + \alpha. \tag{2.17}$$

The fundamental component of the current waveform defined by eqns. (2.15), (2.16) and (2.17) is

$$I = \frac{\sqrt{6}}{\pi} I_d \sqrt{\{[\cos 2\alpha - \cos 2(\alpha + u)]^2 + [2u + \sin 2\alpha -}$$

$$\sin 2(\alpha + u)]^2\}/\{4[\cos \alpha - \cos(\alpha + u)]\}. \tag{2.18}$$

2.7 Invertor operation

The conditions for invertor operation have been described in Section 2.4.1 with reference to an ideal system without commutation reactance. In practice, full inversion cannot be achieved and the delay angle must be less than 180°.

With reference to Fig. 2.11(a) and (e), a commutation from valve 1 to valve 3 (at P3 is only possible as long as phase Y is positive with respect to phase R. Furthermore, the commutation must not only be completed before C6, but some extinction angle γ_1 ($> \gamma_0$) must be left for valve 1, which has just stopped

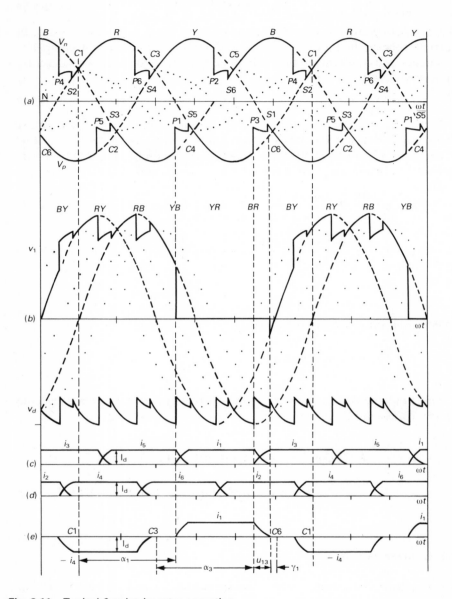

Fig. 2.11 *Typical 6-pulse invertor operation*
    (a) Positive and negative direct voltage with respect to the transformer neutral
    (b) Voltage across valve 1, and direct bridge voltage V_d
    (c), (d) Valve currents i_1 to i_6
    (e) A.C. line current of phase R

conducting, to re-establish its blocking ability. This puts a limit to the maximum angle of firing $\alpha = \pi - (u + \gamma_0)$ for successful invertor operation. If this limit were exceeded, valve 1 would pick up the current again, causing a commutation failure.

Moreover, there is a fundamental difference between rectifier and invertor operations which prevents an optimal firing condition in the latter case. While the rectifier delay angle α can be chosen accurately to satisfy a particular control constraint, the same is not possible with respect to angle γ because of the uncertainty of the overlap angle u. Events taking place after the instant of firing are beyond predictability and, therefore, the minimum extinction angle γ_0 must contain a margin of safety to cope with reasonable uncertainties (values between 15° and 20° are typically used).

The analysis of invertor operation is not different from that of rectification, carried out in Section 2.6, and will not be repeated here. However, for convenience, the invertor equations are often expressed in terms of the angle of advance $\beta\ (= \pi - \alpha)$ or the extinction angle $\gamma\ (= \beta - u)$.

Thus, omitting the negative sign of the invertor d.c. voltage, the following expressions apply:

$$V_d = V_{c0} \cos \gamma - \frac{3X_c}{\pi} I_d \qquad (2.19)$$

or

$$V_d = V_{c0} \cos \beta + \frac{3X_c}{\pi} I_d \qquad (2.20)$$

or

$$V_d = \frac{V_{c0}}{2} (\cos \beta + \cos \gamma). \qquad (2.21)$$

The expression for the direct current is

$$I_d = \frac{V_c}{\sqrt{2}X_c} [\cos \gamma - \cos \beta]. \qquad (2.22)$$

2.8 Power factor and reactive power[4]

Due to the firing delay and commutation angles, the convertor current in each phase always lags its voltage (refer to Fig. 2.10(c)). The rectifier therefore absorbs lagging current (consumes VARs).

34 Static power conversion

In the presence of perfect filters no distorting current flows beyond the filtering point, and the power factor can be approximated by the displacement factor (cos φ), where φ is the phase difference between the fundamental frequency voltage and current components.

Under these idealised conditions, with losses neglected, the active fundamental a.c. power (P) is the same as the d.c. power, i.e.

$$P = \sqrt{3}V_c I \cos \phi = V_d I_d \tag{2.23}$$

and

$$\cos \phi = V_d I_d / (\sqrt{3} V_c I). \tag{2.24}$$

Substituting V_d and I_d from eqns. (2.11) and (2.14) in eqn. (2.24) the following approximate expression results:

$$\cos \phi = \tfrac{1}{2}[\cos \alpha + \cos (\alpha + u)]. \tag{2.25}$$

The reactive power is often expressed in terms of the active power, i.e.

$$Q = P.\tan \phi, \tag{2.26}$$

where tan φ (derived from eqns. (2.18) and (2.24)) is

$$\tan \phi = \frac{\sin (2\alpha + 2u) - \sin 2\alpha - 2u}{\cos 2\alpha - \cos (2\alpha + 2u)}. \tag{2.27}$$

Similarly to eqn. (2.25) the following approximate expression can be written for the power factor of the invertor

$$\cos \phi = \tfrac{1}{2}[\cos \gamma + \cos \beta]. \tag{2.28}$$

Referring to the a.c. voltage and valve current waveforms in Figs. 2.11(*a*) and (*e*) it is clear that the current supplied by the invertor to the a.c. system lags the positive half of the corresponding phase voltage waveform by more than 90°, or leads the negative half of the same voltage by less than 90°. It can either be said that the invertor 'absorbs lagging current' or 'provides leading current', both concepts indicating that the invertor, like the rectifier, acts as a sink of reactive power. This point is made clearer in the vector diagram of Fig. 2.12.

Equations (2.23), (2.25) and (2.26) show that the active and reactive powers of a controlled rectifier vary with the cosine and sine of the control angle respectively. Thus, when operating on constant current, the reactive power demand at low powers ($\phi \simeq 90°$) can be very high.

However such operating condition is prevented in h.v.d.c. convertors by the addition of on-load transformer tap-changers, which try and reduce the steady state control angle (or the extinction angle) to the minimum specified. Under such controlled conditions, Fig. 2.13 shows a typical variation of the reactive power demand versus active power of an h.v.d.c. convertor; the

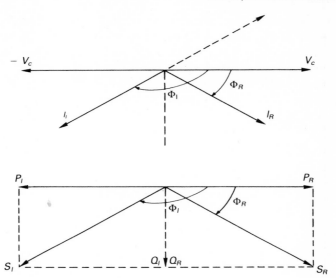

Fig. 2.12 *Vector diagrams of current and power*
  Suffix *R* for rectification
       I for inversion

reactive power demand is shown to be approximately 60 per cent of the power transmitted at full load.

2.9 Convertor harmonics[4]

The term harmonics is used to define the sinusoidal components of a repetitive waveform and these consist exclusively of frequencies which are exact multiples (harmonic orders) of the basic repetition frequency (i.e. the fundamental). The full set of harmonics forms a Fourier series which completely represents the original waveform.

The original waveform can thus be described by its time domain data (i.e. at any given instant in time the amplitude of the waveform is displayed) or by its frequency domain data (i.e. by the magnitudes, and often phase, of its Fourier components).

The general trigonometric form of the Fourier series is

$$F(\omega t) = \frac{A_0}{2} + \sum_{n=1}^{\infty} \{A_n \cos(n\omega t) + B_n \sin(n\omega t)\} \qquad (2.29)$$

where ω is the basic repetition frequency in radians per second, and

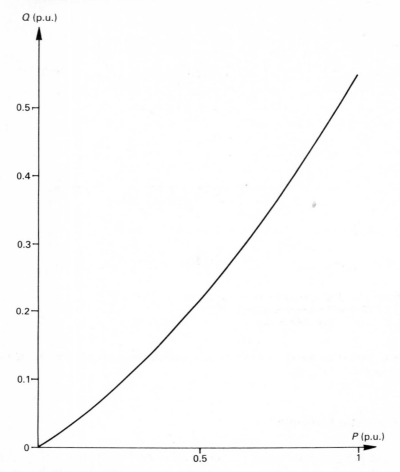

Fig. 2.13 Variation of reactive power with active power

$$A_0 = \frac{1}{\pi} \int_{\sigma}^{\sigma+2\pi} F(\omega t)\, d(\omega t), \tag{2.30}$$

$$A_n = \frac{1}{\pi} \int_{\sigma}^{\sigma+2\pi} F(\omega t) \cos(n\omega t)\, d(\omega t), \tag{2.31}$$

$$B_n = \frac{1}{\pi} \int_{\sigma}^{\sigma+2\pi} F(\omega t) \sin(n\omega t)\, d(\omega t), \tag{2.32}$$

where σ is any angle, $A_0/2$ is the average value of the function $F(\omega t)$ and A_n and B_n are rectangular components of the nth harmonic. The corresponding vector is

$$A_n - jB_n = C_n/\phi_n \tag{2.33}$$

where $C_n = \sqrt{A_n^2 + B_n^2}$ = crest valve and $\phi_n = \tan^{-1}(-B_n/A_n)$.

HVDC convertors generate harmonic voltages and currents on the d.c. and a.c. sides respectively. It is convenient to separate the convertor harmonics in two groups, termed characteristic and non-characteristic, and these are considered in the remaining sections of the chapter.

2.9.1 Characteristic harmonics

The orders of the 'characteristic' harmonics are related to the pulse number of the convertor configuration, defined as the number of non-simultaneous commutations per cycle of the fundamental frequency.

A convertor of pulse number p ideally generates only characteristic voltage harmonics of orders pk on the d.c. side, and current harmonics of orders $pk \pm 1$ on the a.c. side (where k is any integer).

The derivation of the characteristic harmonics is based on the following assumptions:

(a) The supply voltages are displaced exactly by one third of a cycle in time from each other and consist only of fundamental frequency.
(b) The direct current is perfectly constant (i.e. has no frequency components). This can only be achieved if the d.c. smoothing reactor has infinite inductance.
(c) The valves begin conducting at equal time intervals.
(d) The commutation impedances are the same in the three phases (i.e. all the overlap angles are the same).

Direct voltage harmonics: Using as a reference the single three-phase bridge configuration (i.e. $p = 6$) the order of harmonics is $n = 6k$. The repetition interval (see Fig. 2.10(*b*)) is $\pi/3$ and it contains three different functions, which, using as a time reference the voltage crossings, are expressed as follows:

$$V_d = \sqrt{2}V_c \cos\left[\omega t + \frac{\pi}{6}\right] \quad \text{for } 0 < \omega t < \alpha, \tag{2.34}$$

$$V_d = \sqrt{2}V_c \cos\left[\omega t + \frac{\pi}{6}\right] + \frac{1}{2}\sqrt{2}V_c \sin \omega t = \frac{\sqrt{6}}{2} V_c \cdot \cos \omega t$$

$$\text{for } \alpha < \omega t < \alpha + u, \tag{2.35}$$

$$V_d = \sqrt{2}V_c \cos\left[\omega t - \frac{\pi}{6}\right] \quad \text{for } \alpha + u < \omega t < \frac{\pi}{3}, \tag{2.36}$$

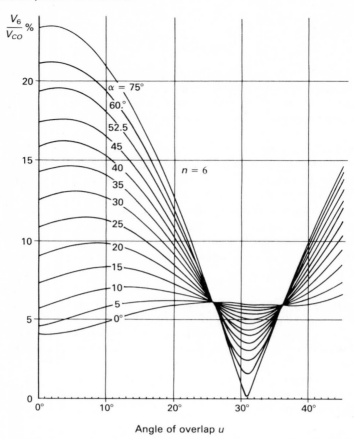

Fig. 2.14 *Variation of 6th harmonic voltage in relation to angle of delay and overlap*

and using the Fourier equations, the rms magnitudes of the harmonic voltages are obtained from the equation

$$V_n = \frac{V_{c0}}{\sqrt{2}(n^2-1)} \left\{ (n-1)^2 \cos^2\left[(n+1)\frac{u}{2}\right] + (n+1)^2 \cos^2\left[(n-1)\frac{u}{2}\right] \right.$$
$$\left. - 2(n-1)(n+1) \cos\left[(n+1)\frac{u}{2}\right] \cos\left[(n-1)\frac{u}{2}\right] \cos(2\alpha+u) \right\}^{1/2}.$$
(2.37)

Figs 2.14 and 2.15 give the 6th and 12th harmonics[2] as a percentage of $V_{c0} = 3(\sqrt{2})V_c/\pi$. These curves and equations show some interesting facts. Firstly for $\alpha = 0$ and $u = 0$, eqn. (2.37) reduces to,

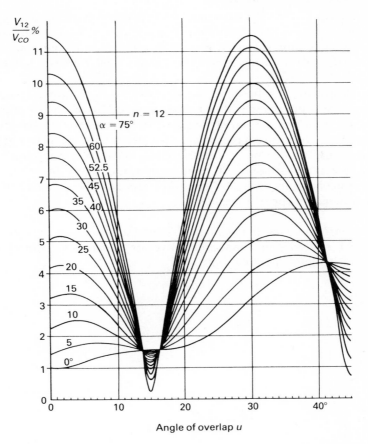

Fig. 2.15 *Variation of 12th harmonic voltage in relation to angle of delay and overlap*

$$V_{n0} = \sqrt{2} V_{c0}/(n^2 - 1) \tag{2.38}$$

or

$$\frac{V_{n0}}{V_{c0}} = \sqrt{2}/(n^2 - 1) \simeq \sqrt{2}/n^2, \tag{2.39}$$

giving 4·04, 0·99 and 0·44 per cent for the 6th, 12th and 18th, harmonics respectively. Generally, as α increases, harmonics increase as well, and for $\alpha = (\pi/2)$ and $u = 0$

$$\frac{V_n}{V_{c0}} = \sqrt{2}\, n/(n^2 - 1) \simeq \sqrt{2}/n, \tag{2.40}$$

which produces n times the harmonics content corresponding to $\alpha = 0$. This means that the higher harmonics increase faster with α. Eqn (2.40) is of some importance as it represents the maximum proportion of harmonics in the system, particularly when it is considered that at $\alpha = 90$ degrees, u is likely to be very small.

If the convertor involves two bridges, one with a star/star connected transformer and the other with a delta/star or star/delta transformer, their voltages will be 30 degrees out of phase and so the harmonics will accordingly be out of phase. Since 30 degrees of main frequency correspond to a half-cycle of 6th harmonic, this harmonic will be in phase opposition in the two bridges. Similarly for 12th harmonic, 30 degrees corresponds to one cycle, giving harmonics in phase; for 18th, 30 degrees corresponds to one and a half cycles, giving harmonics in opposition and so on.

AC current harmonics: In the absence of commutation reactance the current waveform for a star/star connected convertor transformer, shown in Fig. 2.16(a), can be defined as follows:

$$i = I_d \quad \text{for} \quad -\frac{\pi}{3} < \omega t < \frac{\pi}{3},$$

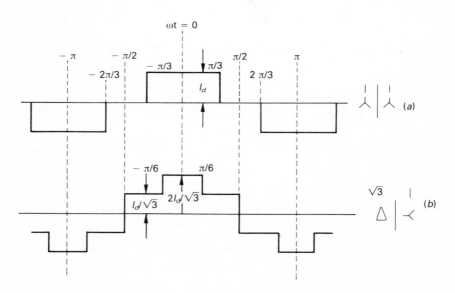

Fig. 2.16 *Idealised phase current waveforms on the primary side*
 (a) Star–star transformer connection
 (b) Delta–star transformer connection

$$i = 0 \quad \text{for} \quad -\frac{2\pi}{3} < \omega t < -\frac{\pi}{3} \quad \text{and} \quad \frac{\pi}{3} < \omega t < \frac{2\pi}{3},$$

$$i = -I_d \quad \text{for} \quad -\pi < \omega t < -\frac{2\pi}{3} \quad \text{and} \quad \frac{2\pi}{3} < \omega t < \pi. \tag{2.41}$$

The Fourier series for such waveform is

$$i = \frac{2\sqrt{3}}{\pi} I_d(\cos \omega t - \tfrac{1}{5} \cos 5\omega t + \tfrac{1}{7} \cos 7\omega t - \tfrac{1}{11} \cos 11\omega t + \ldots) \tag{2.42}$$

with harmonics orders determined from the expression

$$n = 6k \pm 1, \tag{2.43}$$

where k is an integer.

The magnitude of the nth harmonic is given by

$$I_n = \frac{\sqrt{6}}{n\pi} I_d \tag{2.44}$$

and that of the fundamental

$$I_1 = \frac{\sqrt{6}}{\pi} I_d. \tag{2.45}$$

For the star/delta or delta/star transformer connection, the current waveform is given by Fig. 2.16(b) and the Fourier series is

$$i = \frac{2\sqrt{3}}{\pi} I_d(\cos \omega t + \tfrac{1}{5} \cos 5\omega t - \tfrac{1}{7} \cos 7\omega t - \tfrac{1}{11} \cos 11\omega t + \ldots). \tag{2.46}$$

Equations (2.42) and (2.46) are the same excepting that the harmonics 5, 7 (k = odd numbers in eqn. (2.43)) are of opposite sequence, and therefore with two bridges in series as above, only the harmonics corresponding to $n = 12k \pm 1$ will enter the a.c. system.

The current waveform and harmonic spectrum of a double bridge 12-pulse configuration are illustrated in Fig. 2.17 (with the overlap angle ignored).

If the commutation angle is taken into account, the current waveform for the star/star connection has been defined in Section 2.6.2; the characteristic 5th, 7th, 11th and 13th harmonics, as a percentage of the fundamental (I_1), are illustrated in Fig.2.18–2.21 inclusive.[2] It is seen that the harmonics decrease with increases in commutation angle (u), the rate of decrease being greater for higher harmonics. For the same u, changes in α do make little difference outside the range between $\alpha = 0$ and $\alpha = 10$ degrees. A

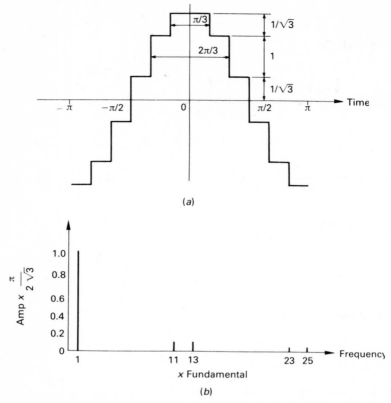

Fig. 2.17 *Idealised phase current waveform with twelve-pulse operation*
(a) Current waveform
(b) Frequency domain representation

simplifying assumption for the analysis of eqn. (2.18) can thus be made by using $\alpha = 0$.

Harmonics tend to reach a minimum at approximately $u = 360°/n$ and then increase slightly. It may also be noted that during normal operation, α is small (say up to 10 degrees) and u is large (say 20 degrees) whereas during disturbances, when α is nearly 90 degrees, u is very small and the harmonics approach their maximum.

For invertor operation α and $(\alpha + u)$ should be replaced by the extinction angle γ and the angle of advance β respectively.

2.9.2 Non-characteristic harmonics

The ideal conditions, used in the last section to calculate the characteristic harmonics produced by h.v.d.c. convertors, are not met in practice and, as a result, relatively small quantities of non-characteristic harmonics are always present. With reference to a.c. current harmonics, the term uncharacteristic

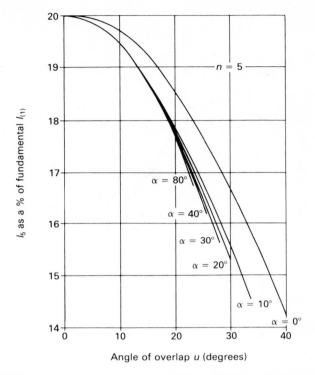

Fig. 2.18 Variation of 5th harmonic current in relation to angle of delay and overlap

indicates frequencies other than those determined by the expression $p.k \pm 1$ (for $k = 0,...,n$) where p is the pulse number of the valve group, i.e. 6 for a single bridge group and 12 for a double bridge group.

Possible causes of non-ideal conditions are:

(a) Firing errors.
(b) AC voltage unbalance (negative sequence) and/or distortion.
(c) Direct current modulation from the remote station.
(d) Unbalance of convertor components.

All these effects cause the convertor to generate non-characteristic harmonics, for example orders 1, 2, 3, etc. on the d.c. side, and 2, 3, 4, etc. on the a.c. side. By way of example, Table 2.1 shows the results of measurements, during back-to-back commissioning tests at the Benmore terminal of the New Zealand scheme.

Effects of firing errors:[4] The Fourier analysis of an ideal 120° rectangular pulse (eqn. (2.42)) shows absence of triplen and even harmonics. However in the event of shorter or longer pulse-widths such harmonics can be generated.

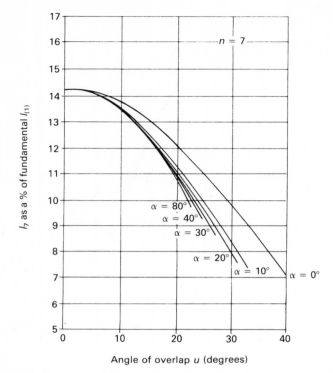

Fig. 2.19 *Variation of 7th harmonic current in relation to angle of delay and overlap*

The ratio of an even harmonic of order n to the fundamental wave at small overlap can be shown to be

$$\frac{I_n}{I_1} = \frac{2 \sin n\varepsilon}{2n \cos \varepsilon} \simeq \varepsilon[1 + \tfrac{1}{3}(n\varepsilon)^2 + \ldots] \simeq \varepsilon \text{ radians}, \tag{2.47}$$

where ε is the angular firing error, such that the valves connected to the common cathode are fired earlier by an angle ε and the valves connected to the common anode are fired late by the same angle.

For example a one degree error, causing a two degrees relative shift between the positive and negative current pulses, will produce approximately 1·74% of second harmonic.

If the firing of the two valves connected to the same phase is late by an angle ε, while the other four valves fire at the right instant, triple harmonics are produced, their ratio to the fundamental being

$$\frac{I_n}{I_1} = \frac{\sin (n\omega/2)}{n \sin (\omega/2)}, \tag{2.48}$$

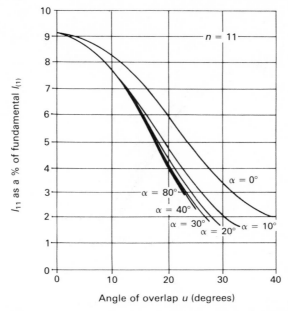

Fig. 2.20 Variation of 11th harmonic current in relation to angle of delay and overlap

where $\omega = 2\pi/3 \pm \varepsilon$ and $n = 3K$. For small errors the above ratio is approximately $\varepsilon/\sqrt{3}$ and therefore a one degree firing error (0·0174 radians) will result in one percent of third harmonic content.[4]

The variation of firing angles from their normal values caused some concern in the early schemes,[5] but with modern thyristor-based schemes under phase-locked oscillator firing control (Chapter 4), this problem has been reduced considerably.

During planning studies the firing error effect is assessed on statistical basis, using a random selection of firing errors of + 0·1, 0·0, − 0·1 for the six valves. The statistical pattern is then used to calculate the harmonic content.

Effect of non-ideal a.c. system voltages:[6] By non-ideal it is meant either the presence of fundamental frequency negative sequence, harmonic distortion, or both. Voltage unbalance produces harmonic content for two reasons. The first reason relates to asymmetrical firing references and its effect has been explained in the last section.

The second reason is the appearance of d.c. current amplitude modulation. The a.c. voltage unbalance produces second harmonic voltage and current components[7] on the d.c. side and these components, shown in Figs. 2.22(*b*) and (*c*) for the case of a 6-pulse convertor, can not be eliminated by equidistant firing control.

46 Static power conversion

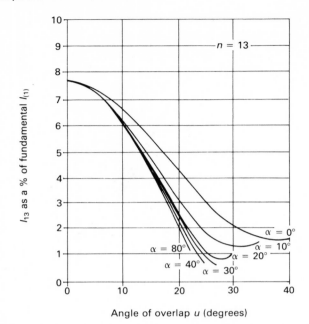

Fig. 2.21 *Variation of 13th harmonic current in relation to angle of delay and overlap*

The phase currents produced by the commutation process on the a.c. side of the convertor are shown in Fig. 2.22(d). The three waveforms are different, but, for each phase, the current waveform is symmetrical, and thus only odd harmonics are present. The Fourier analysis of the phase current waveforms (Fig. 2.22(d)) shows the existence of non-zero sequence triplen harmonics.

Let us now consider the more practical case of a 12-pulse valve group with equidistant firing control. The presence of positive or negative sequence harmonic voltage distortion (of amplitude V_n) per unit of the fundamental frequency produces d.c. voltage harmonic current of order $k = n - 1$ and amplitude V_k. The maximum values of V_k with the commutation reactance neglected, have been investigated by Ainsworth[6] and a summary of the results is illustrated in Table 2.2 for interfering harmonics of orders -5 to $+5$. The amplitude V_k used in Table 2.2 is expressed in per unit of the total V_{c0}, and V_n is expressed in per unit of the nominal a.c. voltage.

The values indicated in Table 2.2 are in practice applicable also to convertors of higher pulse number. Thus, while characteristic harmonics can be reduced by using high pulse number, the non-characteristic harmonics due to a.c. unbalance cannot.

It should be noted that the above results are only an approximation, due to neglect of commutation reactance, but in general they are sufficiently valid at low harmonic orders (< 5).

Table 2.1 *Harmonic measurements during back-to-back testing*

Harmonic	400 A d.c. (one-third full load current) Phase-to-neutral voltages At Benmore on 220 kV		
	Red phase (%)	Yellow phase (%)	Blue phase (%)
1	100	100	100
2	0·5	0·7	1·0
3	2·9	0·3	1·0
4	0·6	0·3	0·4
5	0·25	0·15	0·25
6	0·25	0·30	0·35
7	0·15	0·15	0·1
8	0	0·05	0·1
9	0·05	0·05	0·15
10	0·05	0·05	0·05
11	0·1	0·15	0·1
12	0·15	0·05	0·15
13	0·05	0·05	0·05
14	0·05	0·05	0·05
15	0·15	0	0·2
16	0	0·1	0·15
17	0·3	0·3	0·3
18	0	0·05	0·1
19	0·3	0·3	0·7
20	—	—	—
21	—	—	—
22	0·2	0·2	0·5
23	0·4	0·2	0·3
25	0·2	0·2	0·15

In the case of parallel-connected 12-pulse valve groups the presence of 5th or 7th harmonic distortion on the supply (due for instance to transformer saturation) may cause poor current sharing between the otherwise satisfactory 6-pulse parallel bridges. The sharing can be improved by independent current control of the two bridges, but the delay angles of the two bridges are then different and the injection of '6-pulse' related harmonic content is not eliminated.

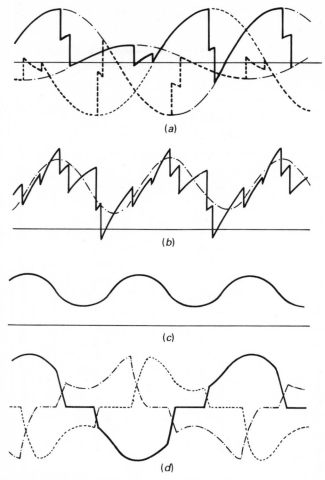

Fig. 2.22 Sustained unbalanced voltages on single convertor bridge
 (a) 3-Phase voltages
 (b) Direct voltage
 (c) Direct current
 (d) 3-Phase currents

Effect of direct current modulation: Let us now consider the case of a perfectly balanced and sinusoidal voltage supply and perfect equidistant firing control. A small amount of d.c. current modulation I_k of harmonic order k will produce a.c. current harmonics I_n of order n.

Neglecting the commutation reactance, the phase sequence and maximum values of I_n for the lower harmonic orders are given in Table 2.3.

The amplitude I_n is expressed in multiples of $I_1 I_k / I_d$ where I_1 = rms fundamental current at the a.c. busbar, I_k = rms interfering current of order k

Table 2.2 *Convertor response to harmonic voltage content on the a.c. supply waveform*

Interfering a.c. voltage harmonic order (n)	Harmonic voltage on d.c. side	
	Order k	Amplitude V_k/V_n
-1	2	0·707
$+2$	1	0·707
-2	3	0·707
$+3$	2	0·707
-3	4	0·707
-4	5	0·707
-5	6	0·771

Table 2.3 *Convertor response to d.c. current harmonic modulation*

Harmonic order (k) of modulating current on d.c. side	Harmonic current on a.c. side	
	Order (n)	Amplitude I_n
1	0	0·707
	$+2$	0·707
2	-1	0·707
	$+3$	0·707
3	-2	0·707
4	-3	0·707
	$+5$	0·707

on the d.c. side, and $I_d =$ d.c. current.

Table 2.3 is independent of the prime cause of the d.c. current modulation. It is again approximate, valid at low frequencies only, because the commutation reactance has been ignored.

Effect of unequal commutation reactances: The maximum values of non-characteristic harmonic current I_n (in per unit of I_1) depend on the mean

Table 2.4 *Effect of unequal commutation reactances for a 0·075 p.u. error between phases*

n	I_n (% of I_1)
3	0·70
5	0·33
7	0·29
9	0·50
11	0·22
13	0·19
15	0·31

commutation reactance X, the commutation reactance error between phases ε and the delay and overlap angles. Typical values are listed in Table 2.4 for a case of $X = 0·2$ p.u., $\varepsilon = 0·075$ and $\alpha = 15°$.

2.10 References

1 UHLMANN, E. (1975): *Power Transmission by Direct Current*, Section 1, Springer-Verlag, Berlin–Heidelberg.
2 ADAMSON, C. and HINGORANI, N. G. (1960): *High Voltage Direct Current Power Transmission*, Chapters 2 and 3, Garraway Ltd., London.
3 ARRILLAGA, J., ARNOLD, C. P., and HARKER, B. J. (1983): *Computer Modelling of Electrical Power Systems*, Chapter 3, John Wiley Ltd., London.
4 KIMBARK, E.W. (1971): *Direct Current Transmission*, Chapters 3 and 8, Wiley-Interscience, New York.
5 REEVE, J. and KRISHNAYYA P. C. S. (1968): 'Unusual current harmonics arising from high voltage d.c. transmission', *Trans. IEEE*, Vol. PAS-87. pp. 883–892.
6 AINSWORTH, J. D. (1981): 'Harmonic instabilities', paper presented at the conference *Harmonics in Power Systems*, UMIST (Manchester).
7 GIESNER, D. B. and ARRILLAGA, J. (1972): 'Behaviour of h.v.d.c. links under unbalanced a.c. fault conditions', *Proc. IEE*, Vol. 119, No. 2, pp. 209–215.
8 ARRILLAGA, J., AL-KHASHALI, H. J. and CAMPOS-BARROS, J. G. (1977): 'General formulation for dynamic studies in power systems including static convertors', *Proc. IEE*, Vol. 124, No. 11, pp. 1047–1052.

Harmonic elimination

Chapter 3

3.1 Introduction

Since the commutation reactance is low in relation to the d.c. smoothing reactance, an h.v.d.c. convertor acts, from the a.c. point of view, as a source of harmonic currents (high internal impedance) and from the d.c. point of view, as a source of harmonic voltage (low internal impedance). The orders and levels of such harmonics have been discussed in Chapter 2.

Excessive levels of harmonic current must be prevented since they will cause voltage distortion, extra losses and overheating, as well as interference with external services (e.g. telephone and railway signals).

The obvious place to eliminate the harmonics is the source itself. In theory, characteristic harmonics could be eliminated either by some complex convertor configuration (which would be uneconomical), or by the use of a series filter preventing the harmonics from arising (which would upset the correct operation of the convertor).

Therefore, accepting that the appearance of harmonics is an inherent property of the static conversion process, it will be necessary to reduce their penetration into the a.c. and d.c. systems.

Any solution which increases the pulse number, reduces the harmonic orders penetrating into both sides of the convertor and should be fully exploited. Beyond the economic range of higher pulse configurations, harmonic elimination will normally require the use of filters.

These are now considered separately.

3.2 Pulse number increase

The relationship between pulse number and harmonic order, discussed in Chapter 2, indicates that the higher the pulse number, the higher the frequency of the lowest order harmonic produced. The use of increased pulse numbers has the following disadvantages:

52   Harmonic elimination

(a) Increased levels of lower order harmonics when convertors are temporarily out of service during maintenance.
(b) Increased number of transformers, both in service and spares.
(c) Increased complexity of transformer connections and the consequent problems of insulation.

Moreover, as the harmonic order increases its amplitude decreases and it is normally cheaper to eliminate it substantially by filtering.

With h.v.d.c. schemes only simple transformer connections are used. This is due to the problem of insulating the transformers so that they withstand the alternating voltages combined with the high direct voltages. A pulse number of twelve is easily obtained with star/star and star/delta transformer connections in parallel, as shown in Fig. 3.1.

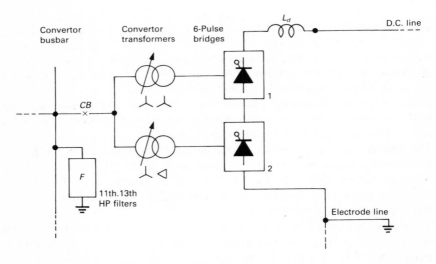

Fig. 3.1   *Twelve-pulse convertor configuration*

3.3 Design of a.c. filters

3.3.1 Design criteria
An ideal in filter design is the elimination of all the detrimental effects caused by waveform distortion, and particularly telephone interference. However, this ideal criterion is unrealistic because of the difficulty of estimating in advance the harmonic flow throughout the a.c. system. It is also uneconomic and, in the case of telephone interference, the problem can normally be solved more economically by taking some of the preventive action in the telephone system itself.

A more practical solution is the reduction of harmonic voltage to an acceptable level at the convertor terminals. The flow of harmonic current causes no special problem provided that the system harmonic impedance is small and therefore a criterion based on harmonic voltage rather than current is more convenient for filter design.

Typical specified factors to be taken into account in filter design are the voltage distortion caused by individual harmonics (V_n), the total voltage distortion defined as

$$V_{TD} = \sqrt{\sum_{n=2}^{\infty} V_n^2}$$

and the telephone influence factor (TIF).

The TIF gives an approximation to the effect of the distorted voltage or current waveform of a power line on telephone noise, without considering the geometrical aspects of coupling. The harmonic frequencies which are sensitive to the ear are given high weighting factors, since even if the harmonic magnitudes are small, these harmonics may result in unacceptable telephone noise. The TIF is defined as

$$\text{TIF} = \frac{1}{V} \left[\sum_{f=0}^{\infty} (K_f P_f V_f)^2 \right]^{1/2},$$

where

$$K_f = 5000(f/1000) = 5f,$$

$$P_f = C\text{-message weighting},$$

$$V_f = \text{rms voltage of frequency } f \text{ on the power line},$$

and

$$V = \left[\sum_{f=0}^{\infty} V_f^2 \right]^{1/2}.$$

3.3.2 Design factors

Two basic concepts in filter design are the filter size and its quality. The size of a filter is defined as the reactive power that the filter supplies at fundamental frequency. It is substantially equal to the fundamental reactive power supplied by the capacitors. The total size of all the branches of a filter is determined by the reactive power requirements of the convertor and by how much this requirement can be more economically supplied by the a.c. generators, extra shunt capacitors, synchronous condensers or static VAR systems (SVS).

The quality of a filter (Q) expresses the sharpness of tuning and is therefore defined differently for tuned and high pass filters. The sharpness of tuning of a resonant filter branch increases with the ratio of its resonance inductance or capacitance to its resistance, whereas in the case of a high pass filter, the sharpness increases in inverse proportion to that ratio.

The high Q, or tuned filter, is sharply tuned to one or two of the lower harmonic frequencies such as the fifth and seventh. The low Q, or damped filter, provides a low impedance over a broad band of frequencies and is often used to eliminate the higher order harmonics, e.g. seventeenth up. It is normally referred to as a high pass filter.

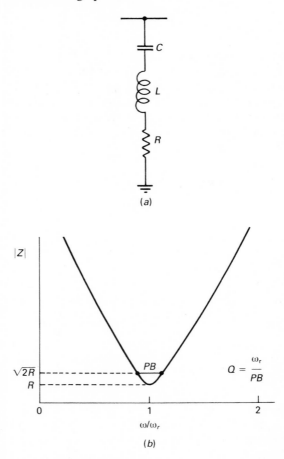

Fig. 3.2 (a) Single-tuned shunt filter circuit
(b) Single-tuned shunt filter impedance

Figures 3.2 and 3.3 show typical circuit diagrams and characterisitics of the two types and Fig. 3.4 illustrates their incorporation within the conventional 6-pulse h.v.d.c. convertor configuration.

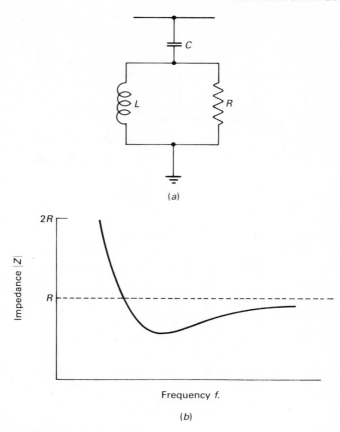

Fig. 3.3 (a) Damped shunt filter circuit
(b) Damped shunt filter impedance versus frequency

The diagram in Fig. 3.5 indicates that the harmonic current generated by the convertor divides between the shunt filters and the a.c. network. To be effective, the filter needs to be of much lower impedance than the a.c. network and ideally must not resonate with the a.c. network impedance.

Therefore the key to good filter design is a clear understanding of the two components of the equivalent circuit, i.e.

(a) the harmonic source (discussed in Chapter 2);
(b) the impedance of the a.c. network at harmonic frequencies.

3.3.3 Network impedance

Kimbark[1] explains that, in general, the harmonic impedance of the a.c. network at the point of filtering will exhibit the following characteristics:

56 Harmonic elimination

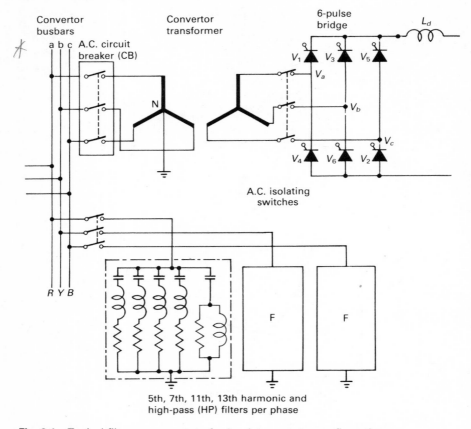

Fig. 3.4 *Typical filter arrangement of a 6-pulse convertor configuration*

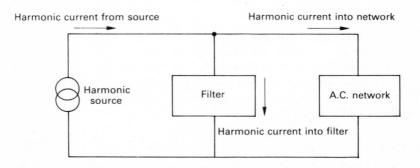

Fig. 3.5 *Simplified circuit of the harmonic source, filter and a.c. network impedance*

(*a*) Alternation of resonance (low resistance) and antiresonance (high resistance) as the frequency increases.

(b) Lower maximum impedance at high loads than at low loads.
(c) Great change in network impedance due to line outages.
(d) Resonances in the a.c. network are the rule rather than the exception.
(e) The harmonic impedances bear no relationship to the fundamental frequency short circuit level.
(f) Loads provide some damping.
(g) This damping increases with frequency.
(h) On cable systems, the impedances to higher order harmonics (15th to 25th) are lower than on overhead line systems.

Figures 3.6 and 3.7 show typical impedance loci in the New Zealand (South Island) 220 kV network for different frequencies and under different operating conditions.

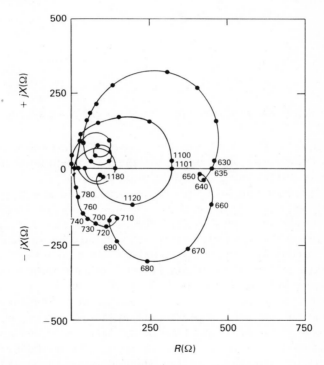

Fig. 3.6 *Harmonic impedances for a strong 220 kV a.c. system*

The impedance loci illustrate the difficulty of estimating the harmonic impedances, even under balanced conditions. Moreover some harmonic effects depend on the harmonic content of the three-phases simultaneously, e.g. communication interference arising from harmonics in the a.c. system is usually caused by the zero sequence components of harmonic currents. If

58 Harmonic elimination

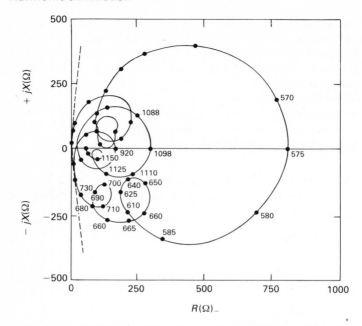

Fig. 3.7 *Harmonic impedances for a weak 220 kV a.c. system*

convertor-generated harmonics are the source of the interference then the zero sequence currents arise solely because of the a.c. system unbalance (i.e. no zero sequence currents are generated by the convertor).

The h.v. transmission lines are often untransposed and this causes the electrical parameters to be different for each phase. Under these conditions the sequence networks are mutually coupled. That is, a current flow of one sequence induces voltages and currents to flow in the other sequences[2].

For very large power ratings it is advisable to use a three-phase equivalent, with the more influential plant components explicitly represented. Such a model is particularly needed when designing filters in the presence of more than one harmonic source. Estimates of fixed current harmonic injections can lead to considerable error, both in the magnitude of the characteristic harmonics and in the order and level of the uncharacteristic harmonics. A multiple harmonic source presents an interesting modelling challenge, involving the sequential use of fundamental power flow and harmonic penetration studies iteratively.

3.3.4 Circuit modelling

Filter design is a complex subject that requires accurate modelling of the behaviour of the harmonic source and of the a.c. system configuration and parameters.

Harmonic elimination

The harmonic source itself, discussed in Chapter 2, is very dependent on the a.c. system conditions. A detailed assessment of the injected harmonic currents requires an iterative procedure involving convertor waveform analysis and harmonic penetration studies. In practice, however, large voltage harmonics are not permissible at the filter busbar and the convertor harmonic content is calculated on the assumption of an infinite busbar source.

Normally computer programmes are used in the calculation of the distortion caused by individual harmonics, the total distortion caused by a given set of harmonic current injections and the telephone influence factor (TIF)

The programmes determine the most critical combination for each harmonic and the corresponding distortion factor. The highest harmonic voltage (i.e. maximum distortion) occurs when the parallel impedance of the filters and a.c. network is maximum (i.e. with parallel resonance). It is, however, unrealistic to expect that parallel resonance occurs at every frequency and only two maximum single distortions are normally considered for the calculation of the total distortion and TIF (i.e. no resonance is assumed for the remaining harmonic currents). Any shunt capacitors present at the convertor terminals must be included in the calculations.

3.3.5 Tuned filters

The early d.c. schemes relied almost entirely on shunt harmonic tuned filters, each of which consisted of a series RLC circuit tuned to the frequency of a low characteristic harmonic. The impedance of a single-tuned filter is

$$Z_f = R + j[\omega L - 1/(\omega C)], \tag{3.1}$$

which at the resonant frequency is a pure resistance R. The pass band of the filter is defined as being bounded by the frequencies at which the filter's reactance and resistance are equal (i.e. when the impedance angle is 45° and its modulus $\sqrt{2}R$).

Figure 3.2 shows a typical impedance curve for the single-tuned filter and the following equations define the filter characteristics:

$$\omega_n = 1/\sqrt{(LC)} = 2\pi f_n \quad (\text{where } f_n \text{ is the resonant frequency}), \tag{3.2}$$

$$X_0 = \omega_n L = 1/(\omega_n C) = \sqrt{(L/C)} = \text{inductive or capacitive reactance at resonance.} \tag{3.3}$$

The quality factor Q can be expressed either as

$$Q = X_0/R \tag{3.4}$$

or

$$Q = \omega_n/PB \quad (\text{where } PB \text{ is the pass band in rad/sec}), \tag{3.5}$$

$$C = 1/(\omega_n RQ), \tag{3.6}$$

$$L = RQ/\omega_n. \tag{3.7}$$

Often two single-tuned filters are replaced by a double-tuned filter. This has proved more economical because it uses only one common inductor and the power loss at fundamental frequency is lower.

In practice a filter is not always tuned exactly to the frequency of the harmonic that it is intended to suppress, for the following reasons:

(a) Variations of the power system frequency, which result in proportional changes in the harmonic frequency.
(b) Changes in the inductance and capacitance of the filter due to ageing and temperature variations.
(c) The accuracy of the actual tuning is restricted by the discrete nature of tuning steps.

The total de-tuning is

$$\delta = \Delta\omega/\omega_n = \Delta f/f_n + \tfrac{1}{2}(\Delta L/L_n + \Delta C/C_n), \tag{3.8}$$

where f_n is the nominal system frequency, L_n is the nominal inductance, and C_n is the nominal capacitance. In terms of δ and Q the filter impedance can be written as

$$Z_f = R[1 + jQ\delta(2+\delta)/(1+\delta)]. \tag{3.9}$$

We are normally interested in small frequency deviations, i.e. $\delta \ll 1$ and therefore:

$$Z_f \simeq R(1 + j2\delta Q). \tag{3.10}$$

Very often, admittances are used instead of impedances, i.e.

$$Y_f = \frac{1}{Z_f} = G_f + jB_f, \tag{3.11}$$

where

$$G_f = 1/[R(1 + 4Q^2\delta^2)], \tag{3.12}$$

$$B_f = -2Q\delta/[R(1 + 4Q^2\delta^2)]. \tag{3.13}$$

The harmonic voltage V_n on the a.c. terminals is then given by

$$V_n = I_n/(Y_f + Y_{sn}) \tag{3.14}$$

or

$$|V_n| = I_n \left\{ \left[G_{sn} + \frac{1}{R(1 + 4Q^2\delta^2)} \right]^2 + \left[B_{sn} - \frac{2Q\delta}{R(1 + 4Q^2\delta^2)} \right]^2 \right\}^{-1/2}, \tag{3.15}$$

where $Y_{sn} = G_{sn} + jB_{sn}$ is the a.c. network admittance at the harmonic frequency 'n'.

The purpose of the filter is to minimise the harmonic voltage given by eqn. (3.15). With reference to eqn. (3.15), the filter variables that can be altered by the designer are the filter size and its quality factor. An optimum filter size is decided by overall cost, and the effect of the filter capacitance on filter cost is illustrated in Fig. 3.8.

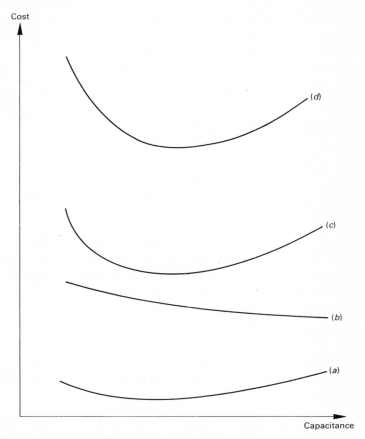

Fig. 3.8 *Relative costs of filter branch versus capacitance*
  (a) Cost of losses
  (b) Cost of inductor
  (c) Cost of capacitor
  (d) Total cost

The filter quality is selected to achieve optimal filter operation, i.e. the selected value should inject minimum harmonic current into the network for the network condition that is assumed. A larger Q reduces the filter losses and the harmonic voltage (when the filter is correctly tuned), but it increases the risk of parallel resonance between the filters and the network.

However, some of the variables in eqn. (3.15) are not under the control of the designer and have to be selected pessimistically, these are:

(a) The frequency deviation δ, which is set to the greatest value that is expected to exist.

(b) The network admittance Y_{sn}. If an accurate specification of system impedances could be made, the filter design would be technically satisfactory and cost effective. This is not normally the case, however, and Y_{sn} is taken as the worst value considered[3], i.e.: (i) very high impedance; (ii) lossless a.c. system; (iii) a.c. system with finite impedance angle.

(i) Very high impedance: In this case it would be possible to choose a very large Q if loss free components could be manufactured. Under these conditions $G_{sn} \simeq B_{sn} \simeq 0$ and eqn. (3.15) reduces to

$$V_n = 2\delta_m X_0 I_n, \tag{3.16}$$

where δ_m is the maximum equivalent frequency deviation, $X_0 = (L/C)^{1/2}$ is the characteristic impedance of the filter resonant circuit, and I_n is the amplitude of the harmonic of order n. In this case the possibility of partial filter/network resonance is disregarded.

(ii) Lossless a.c. system: In this case parallel resonance can occur (i.e. $B_{sn} = -B_f$). Under these conditions eqn. (3.15) reduces to $|V_n| = I_n R(1 + 4Q^2\delta^2)$.

The optimum Q of the filter to minimise the harmonic voltage at detuned conditions is related to the maximum frequency deviation δ_m by

$$Q = 1/(2\delta_m) \tag{3.17}$$

and the maximum harmonic voltage is

$$V_n = 4\delta_m X_0 I_n. \tag{3.18}$$

This case disregards the damping of possible resonance by the real part of network impedance.

(iii) A.C. system with finite impedance angle: The impedance loci of Figs. 3.6 and 3.7 indicate that generally the harmonic impedance can be circumscribed in a part of the plane R, jX determined by two straight lines and a circle passing through the origin. The maximum phase angle of the network impedance can thus be limited to below 90°, and generally decreases with increasing frequency (except in cable networks for high harmonic orders). In this case the highest harmonic voltage is obtained using a phase angle equal to ϕ_{sn} and the opposite sign to that of δ.

Then

$$|V_n| = I_n\{(|Y_{sn}|\cos\phi_{sn} + G_f)^2 + (-|Y_{sn}|\sin\phi_{sn} + B_f)^2\}^{-1/2} \tag{3.19}$$

with ϕ_{sn} positive and δ negative.

Since $|Y_{sn}|$ is unrestricted, the admittance giving maximum $|V_n|$ is

$$|Y_{sn}| = \cos\phi_{sn}(2Q\delta\tan\phi_{sn} - 1)/[R(1 + 4Q^2\delta^2)] \qquad (3.20)$$

giving

$$|V_n| = I_n\omega_n L(1 + 4Q^2\delta^2)/[Q(\sin\phi_{sn} + 2Q\delta\cos\phi_{sn})]. \qquad (3.21)$$

Differentiating eqn. (3.21) with respect to Q, to obtain the lowest harmonic voltage, yields the following expressions for Q and $|V_n|$:

$$Q = (1 + \cos\phi_{sn})/(2\delta\sin\phi_{sn}), \qquad (3.22)$$

$$|V_n| = I_n\delta\omega_n L[4/(1 + \cos\phi_{sn})] = 2I_n R/\sin\phi_{sn}. \qquad (3.23)$$

Nevertheless, it should be noted that filters are not usually designed to give minimum harmonic voltage under these conditions. Normally a higher Q is selected to reduce the losses.

A case that also has to be considered in the design of filters, and which can restrict the operation of the convertors, is an outage of one or more filter branches. The remaining filter branches may then be overstressed, as they have to take the total harmonic current generated by the convertor.

3.3.6 Self-tuned filters[4]

The detuning effects can be compensated by continuous adjustment of the capacitor or inductor (normally the latter), thus permitting the use of high values of Q without having to increase the components ratings. Moreover the resulting low value of the filter resistance increases the filter efficiency.

Self-tuning consists of an on/off servo control which measures the harmonic frequency reactive power in the filter and automatically adjusts the main filter tuning to near resonance, i.e. with filter arm harmonic VARs within the prescribed limits.

The additional cost of providing the inductor variation has to be justified by savings in the capacitor cost and by the improved performance. Whenever the detuning effects are small or when the required fundamental frequency reactive power of the filter is high, self-tuning is not normally an economical alternative.

3.3.7 High pass filters

A high pass damped filter presents a capacitive reactance at the fundamental frequency and a low, predominantly resistive, impedance over a wide band of higher order harmonics.

Since the sharpness of the high pass filter tuning increases with the ratio R/X_0, the Q of this filter normally refers to that ratio (i.e. the inverse of the expression used for the resonant filters). Typical values of Q are between 0·5 and 5.

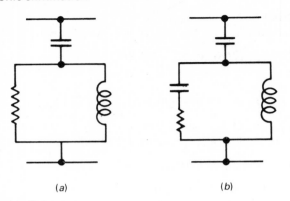

Fig. 3.9 High pass filters
(a) Second order filter
(b) Third order filter

The second and third order high pass filters shown in Fig. 3.9(a) and (b) are extensively used in h.v.d.c. schemes. These are designed to reduce the injection of harmonics above the 17th order into the a.c. system.

When designing such damping filters the Q is chosen to obtain the best characteristic over the required frequency band, and there is no optimal Q as with tuned filters. Because of their bandwidth there is no sensitivity to fundamental frequency deviation or component value drift.

3.3.8 Example of recent filter arrangement

A typical filter/shunt capacitor configuration for a modern 1000 MW convertor station is illustrated in Fig. 3.10(a). The station consists of two 500 MW, 12-pulse groups, one on each side of earth. The filters are divided in two groups, each including 11th and 13th single-tuned arms and a high-pass arm (tuned to the 24th harmonic), as shown in Fig. 3.10(b).

3.3.9 Type C damped filters

With the ratings of some h.v.d.c. links being of the same order as the system short-circuit level, there is an increased probability of low order harmonic resonance between the system impedance and the filter capacitance. Series or parallel resonance will result depending on whether the low harmonic source is within the a.c. system or convertor station respectively.

By way of example, a high probability of third harmonic resonance had been expected on the British side of prospective 2000 MW Cross Channel link. To overcome the problem it was decided to design half the filters for a minimum impedance at around the third harmonic frequency. However, the use of damped filters for low order harmonics involves large fundamental power loss in the damping resistor.

To reduce the power loss of conventional damped filters, a new circuit, illustrated in Fig. 3.11, and called Type C, has been designed[5], where the

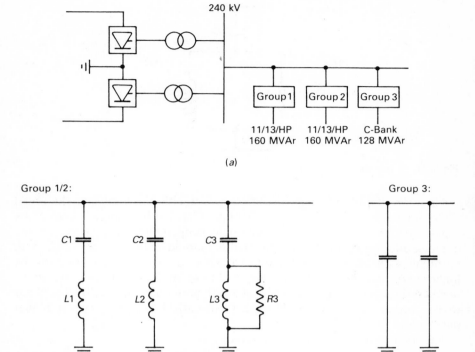

Fig. 3.10 *Filter configuration for a modern 1000 MW convertor station*
    (a) Single line diagram
    (b) Detail of filter and capacitor groups

resistor is bypassed by a fundamental frequency tuned arm (C_2–L). This circuit is more susceptible to frequency variations because of the fundamental frequency tuning but exhibits much lower losses.

3.3.10 Simplified filtering for 12-pulse convertors

Conventional filter design, based on the use of separate tuned filters of the series resonant type for the 11th and 13th harmonics and a high pass filter for the higher order harmonics, provides a more effective reduction of harmonics than it is normally required. In the conventional design the minimum size of the filters is usually determined by the available economic size of capacitor units and by the minimum amount of reactive power compensation required to be provided at the convertor's terminal.

Harmonic elimination

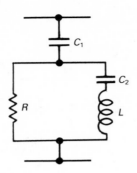

Fig. 3.11 C-type band pass filter

Therefore the filter design can be simplified, either by replacing the tuned filters for harmonics 11 and 13 by a single filter of the damped type, or by replacing all the individual filters by a single damped filter. In the first case, the damped filter replacing the two tuned filters should be tuned to about the 12th harmonic with a fairly high Q (20 to 50) while the damped filter used for the higher harmonics has a much smaller Q (2 to 4). In the second case, the single damped filter is also tuned to about the 12th harmonic, but a fairly low Q has to be chosen (2 to 6) to achieve a sufficiently low impedance at higher harmonics.

The advantages of the damped filter are:

(a) the performance and loading are less sensitive to temperature, system frequency deviations and component tolerances;
(b) because a wide spectrum of harmonic frequencies is filtered, the considerable cost of subdividing the filter into several separate arms is avoided; this also leads to a reduced site area;
(c) maintenance is reduced;
(d) uncharacteristic harmonics are also absorbed, subject to the filter Q and centre frequency;
(e) the need to carry out tuning on site is reduced or eliminated;
(f) it is easier and cheaper to split the filter into smaller subgroups for reactive power control; sharing the harmonic current between these subgroups present no problem.

On the other hand, damped filters need to be bigger in terms of fundamental MVAR to achieve the same level of filtering performance as tuned resonant filters. The harmonic losses in tuned resonant filters are usually lower than in damped filters, while the opposite is true for the fundamental frequency losses.

3.4 D.C. side filters

On the d.c. side of h.v.d.c. convertors the voltage harmonics generate

Harmonic elimination

harmonic currents, whose amplitudes depend on known elements such as the delay and extinction angles, the overlap angle, the impedance of d.c. circuits (i.e. smoothing reactors, damping circuits, surge capacitors and the line itself).

In the case of overhead line transmission, telephone interference can be very substantial. Therefore comprehensive studies must be carried out at the planning stage, in order to decide whether to use filters or reroute parts of the transmission lines away from telephone systems.

The assessment of interference levels requires detailed information of the harmonic voltage and current profiles along the h.v.d.c. line; electromagnetic induction from harmonic currents is normally the main problem. Moreover, since both ends of the link contribute to the disturbance, it is necessary to obtain the profiles from each end and add their effects.

Computer programmes have been developed for this purpose[6] which calculate the self impedance, mutual impedance and capacitive matrices for any frequency and line configuration. This information is then used to evaluate the generalised transmission line equations at defined intervals along the line section. Considering the convertor as a harmonic voltage source at each end, the programmes calculate the sending and receiving end currents and thus the individual current and voltage profiles at specific locations.

As an estimated worst case, the results are then combined by adding the root of the sum of squares (of the peak current profile) derived from each end of the link. At each harmonic a profile along a single equivalent conductor is obtained. This is determined from the vector addition of the harmonic current values from each of the d.c. conductors and the d.c. overhead ground wire. The equivalent conductor is assumed to be located along the centre of the d.c. lines. A typical 12-harmonic current profile, with filters included, is illustrated in Fig. 3.12.

Three different criteria have been used to define the need for and the performance of d.c. filters in d.c. transmission schemes.

(a) Maximum voltage TIF (Telephone Influence Factor) on the d.c. high voltage bus.
(b) Maximum permissible noise to ground in telephone lines close to the h.v.d.c. line.
(c) Maximum induced noise intensity in a parallel test line one kilometer away from the h.v.d.c. line.

Typical types and location of d.c. filters in several existing schemes are shown in Fig. 3.13, and the reasons for their selection are discussed in Reference 7.

Component ratings are considerably different to those for an a.c. filter, since the harmonic current is reduced to a relatively small value by the large d.c. smoothing reactor; consequently the capacitor cost is almost entirely dependent on its capacitance and the d.c. voltage.

68   Harmonic elimination

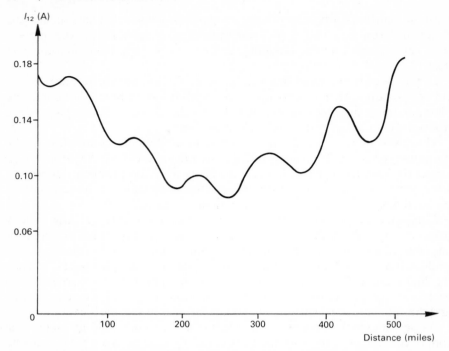

Fig. 3.12   *Twelfth-harmonic current profile along a d.c. transmission line*

Similarly to the a.c. filter, the Q is selected using eqn. (3.22). Being the most expensive component, the capacitor is chosen first and then, for a given frequency, the value of the inductor can be determined.

Under extreme conditions of telephone interference, the inductance of the d.c. reactor may be increased, or two reactors may be connected in tandem, in which case the shunt branches should be connected to the point between the two reactors.

3.5 Alternative methods of harmonic elimination

Because of the complexity and costs of filters there have been several attempts to achieve harmonic control by other means, i.e.:

(*a*) elimination by magnetic flux compensation;
(*b*) elimination by harmonic injection;
(*c*) elimination by d.c. ripple reinjection.

3.5.1 *Magnetic flux compensation*

This method of harmonic elimination is basically illustrated in Fig. 3.14. A current transformer is used to detect the harmonic components coming from the non-linear load. These are fed, through an amplifier, into the tertiary

Harmonic elimination 69

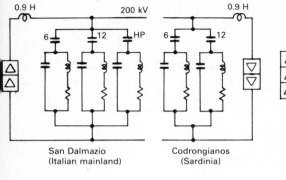

(a) Sardinia

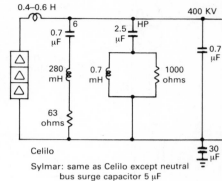

(b) Pacific intertie

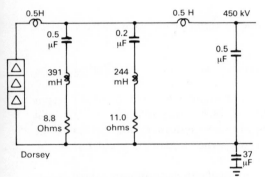

(c) Nelson River Bipole 1

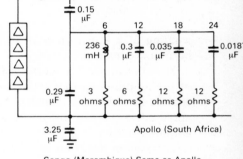

(d) Cabora Bassa

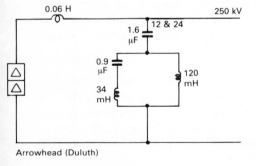

(e) Square Butte

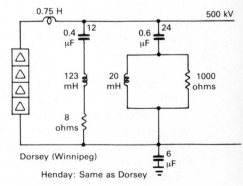

(f) Nelson River Bipole 2

Fig. 3.13 D.C. filter circuits of various h.v.d.c. schemes (from Reference 7)

70 Harmonic elimination

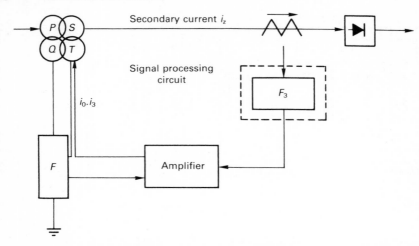

Fig. 3.14 Basic configuration of the harmonic current elimination method by flux compensation

winding of a transformer in such a manner as to cause cancellation of the harmonic currents concerned.

The main areas of concern with this system involve the coupling of the output of the amplifier to the tertiary winding in such a way that the fundamental current flow does not damage the amplifier. A quaternary winding and filter are used, as shown in Fig. 3.14, to reduce the fundamental current in the amplifier output.

The system has been developed to an advanced experimental stage and is discussed fully in an IEEE paper[8]. Typically the amplifier rating is in the order of 750 KW for 300 MW rectifier loads.

One advantage with this scheme is its ability to take account of uncharacteristic harmonics, such as the third and ninth. The main disadvantage of the scheme is its inability to effectively remove the large magnitude lower order characteristic harmonics without the need for a very high power feedback amplifier.

3.5.2 Harmonic injection

Another means by which harmonics can be eliminated is to modify the convertor rectangular current waveform by adding a harmonic current from an external source, as shown in Fig. 3.15. In such a scheme, as originally proposed by Bird[9] and further developed by Ametani[10], a triplen harmonic from the source is injected in the conducting transformer phases.

The advantage of these schemes over filtering is that the system impedance is not part of the design criteria. However they suffer from the following disadvantages:

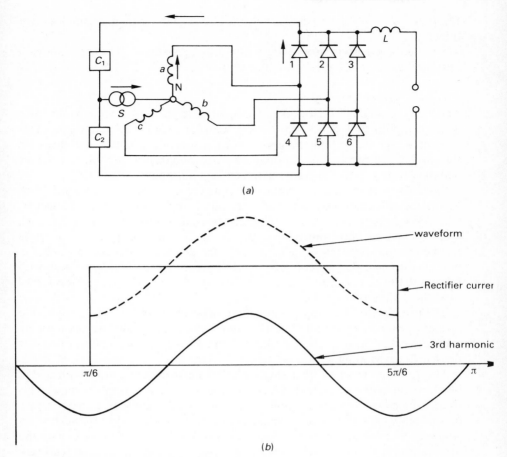

Fig. 3.15 *Harmonic injection technique*
    (a) Circuit for a bridge rectifier
        L = smoothing reactor
        N = neutral point of d.c. windings
        S = injected current source
        C = cutoff circuit of direct current
    (b) Current waveforms

(*a*) need of a triplen harmonic current generator and its synchronisation to the supply main frequency;
(*b*) difficulty in adjusting the amplitude of the sinusoidal injected current to suit each particular operating condition;
(*c*) difficulty in adjusting the phase of the sinusoidal injected current to suit each particular operating condition;
(*d*) inability to nullify more than one harmonic order at any operating point;

(e) poor efficiency due to ineffective dissipation of the triplen harmonic power injected;
(f) the passive system proposed by Bird although solving (a) and (b) above is only applicable to rectifiers operating with 0° delay.

3.5.3 D.C. ripple injection[11]

The basic principle consists of developing a triple frequency current-wave whose amplitude depends on the d.c. current magnitude and shape. This current is injected into the main transformer secondary neutral and flows via the conducting transformer winding. The modified transformer phase currents will then contain only 12-pulse related current harmonics.

Static convertor schemes produce a ripple voltage at their d.c. output. With 6-pulse rectification the ripple has a period of $1/6T$ where $T = 1/f$. However, with respect to the star point of the convertor side transformer windings, each d.c. pole has a non-sinusoidal ripple voltage of period $1/3T$, i.e. a triple frequency voltage. This voltage has the same phase relationship on each d.c. pole, in the sense that the common anode and common cathode potentials with respect to the neutral both increase or decrease simultaneously, and is referred to as the common mode d.c. ripple voltage.

The principle of d.c. ripple reinjection, illustrated in Fig. 3.16, is applicable to static convertors with 120° conducting valves. The transformers must be star-connected on the rectifier side and must have either a delta primary or delta tertiary winding.

The primary winding of a single-phase transformer in series with a d.c. blocking capacitor is connected to the common mode d.c. ripple voltage. This transformer provides the commutating voltage for a single phase triple-frequency full-wave rectifier (or feedback convertor) connected to the secondary winding. The output of the feedback convertor is connected in series with the d.c. output of the 6-pulse convertor. The a.c. output of the feedback transformer is therefore a triple frequency square wave current, which is then adjusted to the appropriate level by the transformer ratio. The primary triple-frequency injected current is illustrated in Fig. 3.17(a).

The d.c. ripple of the 6-pulse convertor and the feedback convertor combine to produce a 12-pulse waveform on the d.c. side secondary (or convertor side) of the main transformer, as shown in Fig. 3.17(e). This applies equally for a purely resistive load.

The frequency of the harmonic injection is derived from the supply frequency and, therefore, the problem of synchronising the harmonic source with the mains frequency does not arise. Moreover, the problem of injected-current phase adjustment is solved by using controlled-rectifier feedback. The firing angle control of the feedback convertor is thus locked to the main rectifier control, i.e. if the thyristors of the feedback convertor are fired 30° after the corresponding main convertor thyristors then the waveforms illustrated in Fig. 3.17 result.

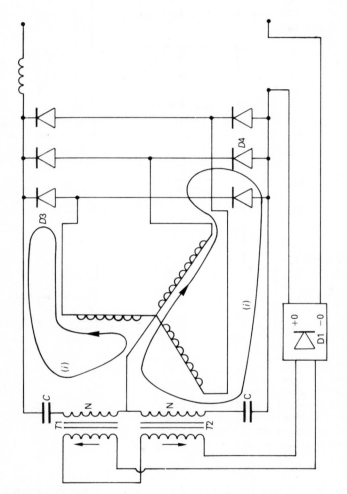

Fig. 3.16 *Bridge rectifier with ripple reinjection*
T1,T2 Feedback rectifier transformers $N:1$ turns ratio
C Blocking capacitor
D1 Feedback rectifier
(i) Path of injected current while *D3* and *D4* on

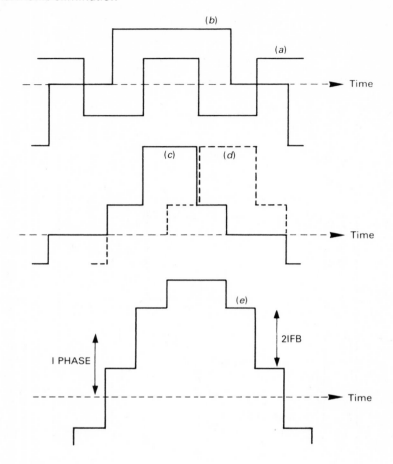

Fig. 3.17 *Inductive d.c. load*
  (a) Triple frequency injected current
  (b) Rectifier current before modification
  (c) Modified phase current, rectifier winding
  (d) Second phase displaced 120°
  (e) Resultant phase current on delta primary

A Fourier analysis of waveform 3.17(e) shows that for a particular ratio of the injected-to-rectifier current ratio all harmonics of order $6n \pm 1$ (where $n = 1, 3, 5, \ldots$) are zero while the other harmonic orders (i.e. for $n = 2, 4, 6, \ldots$) retain the same relationship with the fundamental as before. The result is that the original 6-pulse convertor configuration has been converted into a 12-pulse convertor system, from the point of view of a.c. and d.c. system harmonics.

2 KUUSSAARI, M. and PESONEN, A. J. (1976): 'Measured power line harmonic currents and induced telephone noise interference with special reference to statistical approach', *CIGRE, Paper 36–05*, Paris.
3 CORY, B. J. (Editor (1965): *High Voltage Direct Current Convertors and Systems*, p. 151, MacDonald, London.
4 CLARKE, C. D. and JOHANSON-BROWN, M. J. (1966): 'The application of self-tuned harmonic filters to h.v.d.c. convertors', *IEE Conference on High Voltage D.C. Transmission*, Publication 22, pp. 275–276.
5 STANLEY, C. H., PRICE, J. J., and BREWER, G. L. (1977): 'Design and performance of a.c. filters for 12-pulse h.v.d.c. schemes', *IEE Conference Publication 154 on Power Electronics-Power Semiconductors and Their Applications*.
6 OUELETTE, K. R. and LEWIS, D. W. (1971): 'Harmonic interference from d.c. lines', *Manitoba Power Conference EHV-DC*, Winnipeg, pp. 543–578.
7 HARRISON, R. E. and KRISHNAYYA, P. C. S. (1978): 'System considerations in the application of d.c. filters for h.v.d.c. transmission', *CIGRE Paper 14–09*, Paris.
8 SASAKI, H. and MACHIDA, T. (1971): 'A new method to eliminate a.c. harmonic currents by magnetic flux compensation. Considerations on basic design', *Trans. IEEE*, Vol. PAS-90, pp. 2009–2019.
9 BIRD, B. M. *et al.* (1969): 'Harmonic reduction in multiplex convertors by triple frequency current injection', *Proc. IEE*, Vol. 116, pp. 1730–1734.
10 AMETANI, A. (1972): 'Generalised method of harmonic reduction on a.c./d.c. convertors by harmonic current injection', *Proc. IEE*, Vol. 119, pp. 857–864.
11 BAIRD, J. F. and ARRILLAGA, J. (1980): 'Harmonic reduction in d.c. ripple reinjection', *Proc. IEE*, Vol. 127, No. 5, pp. 294–303.

Chapter 4
Control of h.v.d.c. convertors and systems

A-CONVERTOR CONTROL

4.1 Basic philosophy

The ideal control system for an h.v.d.c. convertor should meet the following requirements[1]:

(*a*) Symmetrical firing of the valves under steady state conditions.
(*b*) Instant of firing to be decided with regard to permissible values of commutation voltage (rectifier), and commutation margin (invertor).
(*c*) Minimal reactive power consumption in the convertors, subject to the condition that it is achieved without an unacceptable risk of commutation failure.
(*d*) Insensitivity to normal variations in voltage and frequency of the a.c. supply network.
(*e*) Some degree of prediction of the optimum instant of firing in the invertors, based on actual network voltage and direct current, subject to the condition that it is achieved without an unacceptable risk of commutation failure.
(*f*) Current control characteristics with sufficient speed and stability margin to cope with changing reference values and disturbances.
(*g*) Continuous operating range from full rectification to full inversion.

The theory presented in Section 2.5 is based on perfectly symmetrical and sinusoidal waveforms with the firing angles (α) occurring at exactly equal intervals and in the appropriate cyclic sequence. Deviations from such ideal conditions give rise to two basically different control methods which are discussed in Sections 4.2 and 4.3.

4.2 Individual phase-control

This is the method used in the early h.v.d.c. convertors. The firing instants are determined individually for each valve, so that a constant delay (or extinction) angle is maintained for all the valves in the steady state with respect to the earliest firing instant (i.e. the voltage crossing).

In rectification the constant delay angle is normally determined from a negative feedback control loop, involving the set current and the actual monitored current, as shown in Fig. 4.1.

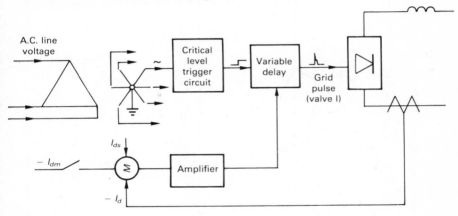

Fig. 4.1 *Constant-α current control system*

To maintain safe invertor operation with minimum reactive power requirements the individual firings require:

(*a*) a continuous calculation of the available voltage-integral for commutation;
(*b*) a continuous calculation of the required voltage-integral for safe commutation.

Optimum firing is achieved when the results of (*a*) and (*b*) coincide.

The relationship governing the commutation process relies on the fact that the time integral of the commutating voltage, i.e. the voltage integral, is equal to the overall voltage change produced by the commutating current i_c. With reference to Fig. 4.2 such relationship can be expressed as

$$\int_\alpha^{\pi-\gamma_0} \sqrt{2} V_c \sin(\omega t)\, d(\omega t) = 2(\omega L) \int_0^{I_d} di_c$$

or

$$\sqrt{2} V_c \cos \alpha + \sqrt{2} V_c \cos \gamma_0 = 2\omega L I_d.$$

Moreover, during large or small system disturbances the actual current at the end of the commutation period will be different from the magnitude

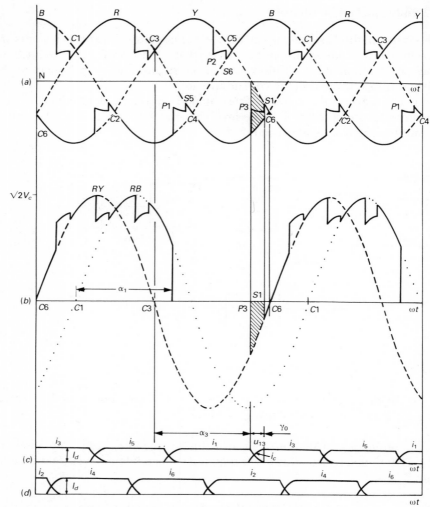

Fig. 4.2 *Typical 6-pulse invertor under predictive CEA control*
 (a) Positive and negative direct voltages with respect to the transformer neutral
 (b) Voltage across valve 1, and commutating voltages
 (c) Valve current i1, i3, and i5
 (d) Valve current i2, i4, and i6

anticipated by the controller, and compensation is made for the rate of change of current. Thus the equation used as a basis for a 'predictive' constant extinction angle (CEA) controller is[2]

$$\sqrt{2}V_c \cos \alpha + \sqrt{2}V_c \cos \gamma_0 = 2\omega L\left(I_d + u\frac{dI_d}{d\omega t}\right), \quad (4.1)$$

where α is the firing delay angle, γ_0 is the preset CEA, u is the commutation angle and V_c is the commutating voltage (rms), e.g. $V_c(RB)$ is used to compute eqn. (4.1) for valve 1.

This type of control has the advantage of being able to achieve the highest direct voltage possible under asymmetrical or distorted supply waveforms. On the other hand any deviations from the ideal voltage waveforms will break the 120° symmetry of the current waveform and thus cause extra waveform distortion, as explained in Section 2.9.2.

Operating difficulties encountered in the early schemes indicated that the distorted convertor currents cause a.c. voltage distortion which can influence the firing pulse spacing via the control system, and often reinforce the original distortion.

In theory this situation can be improved by placing control filters between the a.c. system and the control system, so as to attenuate the harmonics. However the use of control filters has disadvantages too, like the filter's inherent phase error, which varies with system frequency, and its inability to attenuate negative sequence fundamental voltages, whose effect is precisely to cause irregular firing-pulse spacings. Moreover with filters included, the control system will ignore the presence of harmonics on the a.c. voltages, whereas the valves will respond to the actual voltages reaching them, including harmonics.

4.3 Equidistant firing control

The difficulties encountered with the original scheme encouraged the development of an alternative control philosophy which could get away from the voltage waveform dependence. A new principle, initially referred to as the 'phase-locked oscillator', appeared in the late 1960s[3] which largely achieved the target.

The basis of this control system, illustrated in Fig. 4.3, is a voltage-controlled oscillator which delivers a train of pulses at a frequency directly proportional to a d.c. control voltage V_c.

The train of pulses is fed to a six-stage ring-counter in which only one stage is on at a time; the ON stage is stepped cyclically from positions 1 to 6 by the oscillator pulses. As each ring-counter stage turns on, it produces a short pulse at the output (once per cycle). Therefore the complete set of six output pulses normally occur as successive intervals of 60°. The STOP pulses are also obtained from the ring-counter but two stages later (e.g. the START pulse for valve 1 is from stage 1 and the STOP pulse for valve 1 is from stage 3, normally 120° later). One oscillator and one ring-counter per bridge constitute the basic control hardware.

The various control modes only differ in the types of control loop which provides the oscillator control voltage V_c.

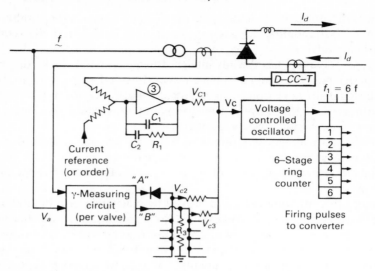

Fig. 4.3 *Principle of the phase-locked oscillator control system*

The phase of each firing pulse will have some arbitrary value relative to the a.c. line voltages, i.e. an arbitrary value of convertor firing angle α. However, when the three-phase a.c. line voltages are symmetrical fundamental sine waves, α is the same for each valve.

In practice the simple independent oscillator would drift in frequency and phase relative to the a.c. system; hence some method of phase-locking the oscillator to the a.c. system is required. This is normally achieved by connecting V_c in a conventional negative feedback loop for constant current or constant extinction angle, as described in the following two sections.

4.3.1 Constant current loop

With reference to Fig. 4.3, let us first consider the constant current loop, i.e. only signal V_{c1} being effective. This voltage is obtained from the amplified difference (error) between the current reference and the measured d.c. line current; this forms a simple negative-feedback control loop, tending to hold current constant at a value very close to the reference.

To visualise the operation of this loop, imagine that the current is nearly equal to the reference, such that the amplified error (V_c) happens to be precisely that value required to give an oscillator frequency of six times the supply frequency. The ring-counter outputs, and thus the valve-gate pulses, will have a certain phase with respect to the a.c. system voltage. Suppose further that this phase, which is identical to firing angle α, happens to be such as to give the correct convertor d.c. output voltage, which, with the particular back e.m.f. of the d.c. link, results in the correct d.c. line current. This is steady state operation.

The loop is self-correcting against disturbances of any source. For instance, a drop of back e.m.f. in the d.c. system causes a temporary current increase, which reduces V_c and hence slows down the oscillator, thus retarding its phase and finally increasing the firing angle α. This tends to decrease the current again, and the system settles down to the same current, with the same V_c and oscillator frequency but a different phase, i.e. different α.

The control system will also follow system frequency variations, in which case the oscillator has to change its frequency; this results in different V_c and hence current, but the current error is made small by using high gain amplification.

This constant current scheme is the main control mode during rectification; it is also used during inversion whenever the invertor has to take over the current control, as explained in Section 4.7.

The control system response is fast but, in practice, its effect will be slowed down by the relative slower response of the d.c. line which includes capacitance, inductance and smoothing reactance.

4.3.2 Invertor extinction angle control

This control mode is implemented by a negative feedback loop very similar to the current loop and is also shown in Fig. 4.3. The difference between the measured γ and the γ-setting is amplified and provides V_c as before. However it differs in that γ is a sampled quantity rather than a continuous quantity. For each valve the extinction angle is defined as the time difference between the instant of current zero and the instant when the anode voltage next crosses zero, going positive. Typical waveforms of the γ-measuring technique are shown in Fig. 4.4.

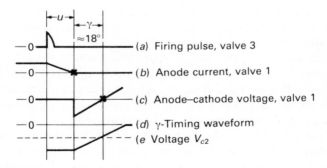

Fig. 4.4 Waveforms of γ-control circuit

For each bridge there are six values of γ to be measured, which under symmetrical steady state operation are identical. Under unbalanced conditions, however, the valve with greatest risk of commutation failure is the one having the smallest γ; this smallest measured γ produces the most negative output and thus causes its diode (signal A in Fig. 4.3) to conduct and produce the negative feedback voltage V_{c2}. During steady state operation and full inversion, V_{c2} controls the oscillator holding the smallest γ at a predetermined value by closed-loop control.

Under these conditions V_{c1} is zero because the invertor C.C. setting is less than the d.c. line current (determined by the rectifier C.C. control); hence the invertor C.C. loop is trying to decrease α by making V_{c1} as low as possible. The minimum V_{c1} is clamped to zero volts and thus during normal invertor operation the C.C. loop if ineffective.

Component B is an additional feedback voltage (V_{c3}) applied during the transient condition when γ < $γ_{min}$. A sudden impulse is then applied to the voltage controlled oscillator, which has an integrating characteristic and thus can suddenly shift the phase (i.e. angle α) by an appropriate amount.

4.3.3 Transition from extinction angle to current control

The current setting for invertor operation is below that of the current in the d.c. line, which is normally kept at a higher constant level by the sending end terminal (explained in Section 4.7). Under such conditions V_{c1} in Fig. 4.3 is zero, the current control amplifier is saturated and the convertor operates on constant extinction angle γ at full inversion.

In the event of a sudden a.c. system voltage rise at the invertor end, or a d.c. line voltage reduction, the direct current will decrease; the current amplifier then comes out of saturation and V_{c1} becomes positive. This additional input to the oscillator causes operation of a larger γ, i.e. advances the firings and the convertor takes over current control.

4.3.4 Other equidistant firing control schemes

Various versions of the original phase-locked oscillator control system have subsequently appeared.

The basic principle of an early alternative[4] is illustrated in Fig. 4.5, where a control function V_{cf} of linear slope initiates the firing signal at the intersection with the controller output voltage V_c. At that instant V_{cf} returns to its initial value and starts again. The time between consecutive impulses is therefore determined by the magnitude of V_c and the slope of V_{cf}.

Let us assume that the value of V_c in Fig. 4.3 has been selected so that the impulse interval is precisely 60° of the actual system frequency, and that the position of the impulse corresponds with a particular delay angle α. If V_c remains unchanged, the sequence of firing instants is determined by the dotted vertical lines. With a temporary increase in V_c the impulse spacing increases and the firing sequence is determined by the full vertical lines. For each impulse occurring while V_c is higher, the delay angle is increased by the amount Δα. Following the return to the original V_c the total increase (3Δα in the illustration) is retained.

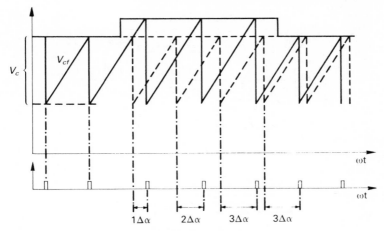

Fig. 4.5 *Firing pulses of alternative equidistant firing control*

Some schemes involve an element of prediction of the extinction angle, similar to the conventional individual phase-control scheme; the prediction is made using the latest values of a.c. voltage and d.c. current in the commutation equation, as described in Section 4.2.

One of these versions, entitled 'Equidistant firing predictive-type control'[5], claims that the prediction is effective for the incoming firing, and uses a feedback loop to update the predictive model for the subsequent firings.

In both the basic phase-locked oscillator and the equidistant firing predictive schemes, a change in the control voltage changes directly the frequency of the oscillator, and the synchronisation of the oscillator takes place with the help of the main current control loop or the extinction angle control loop.

In an alternative equidistant firing control scheme,[6] illustrated in Fig. 4.6, a voltage controlled current source charges the capacitor C. When the voltage across this capacitor exceeds a given control voltage V_{c1} by an amount $\Delta V/2$, the voltage comparator produces an output pulse and the capacitor is rapidly discharged to the voltage $V_{c1} - \Delta V/2$; the charging of the capacitor then begins once again. The output pulses of the voltage comparator are distributed to the convertor valves via a shift-register.

In Fig. 4.6 the frequency of the oscillator is determined by V_{c2} which results from the sum of two signals V_{c21} and V_{c22}. V_{c21} is proportional to the a.c. system frequency and V_{c22} is the output of a s ow acting α-control loop. The phase of the control pulses is determined by the control voltage V_{c1} which is derived (see Fig. 4.7) either from a proportional-integral acting current controller or from a proportional-integral acting extinction angle controller. The $\cos^{-1}$ circuit compensates for the non-linear characteristic of the convertor. To ensure invertor stability it is important to keep negative deviations of the extinction angle as small as possible, and the

84 Control of h.v.d.c. convertors and systems

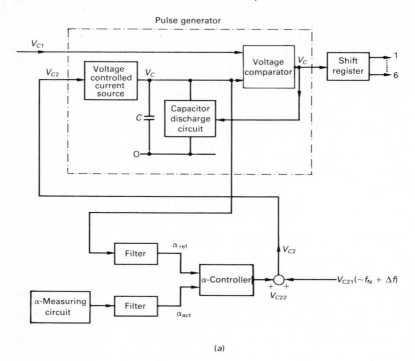

(a)

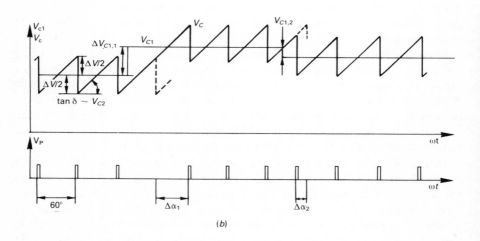

(b)

Fig. 4.6 *Pulse phase grid control system (PPC) (© 1970 IEEE)*
  (a) Block diagram
  (b) Determination of the phase of the firing pulses

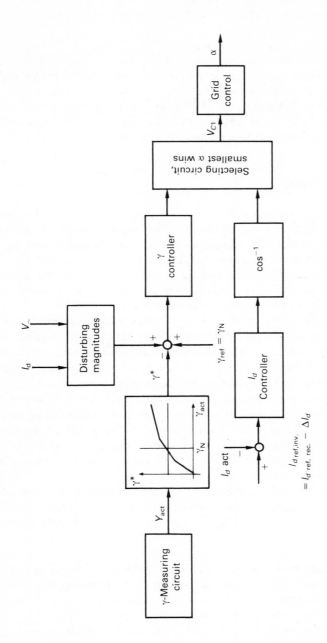

Fig. 4.7 *Block diagram of the closed-loop extinction angle control with asymmetrical controller and of the current control with selecting circuit (© 1970 IEEE)*

measured γ is presented to the controller through a shaping network. The additional disturbing magnitudes provide the element of prediction.

4.3.5 Application to 12-pulse convertor groups
When bridges are connected in series, the common direct current is determined by one controller per pole and the control pulses are directed to the individual bridges from a common ring-counter. With 12-pulse convertor groups it is thus necessary to double the number of control impulses per cycle. In this case the ring-counter provides the control pulses to the twelve different valves of the two bridges, keeping four valves activated simultaneously.

4.4 Comparative merits

For the normal working range, equidistant firing control is preferable to individual phase control because of the reduction of abnormal harmonics. As with any control system, its firing pulse spacing may be modulated by any non-characteristic harmonics entering via feedbacks, e.g. of d.c. current. Thus its description as 'equidistant firing control' is somewhat anomalous; however the phase-locked oscillator control system is very stable when suitably designed, and this is not usually an important effect.

It is often claimed that, whenever the d.c. power contributes considerably to maintaining the network stability, such system may benefit from a change-over to individual phase control under asymmetrical network fault conditions. However under asymmetries the power transmitted is limited by the current (with individual control) or the voltage (with equidistant control) and therefore the above conclusion must be restricted to asymmetries which require no current limiting. For large asymmetries, with the power decreasing to very small levels (with either control), the reduction of direct voltage during the fault is less important; in such cases the use of equidistant firing control provides more reliable commutations and facilitates the rapid return of power flow when the asymmetry ends.

On the subject of adding predictive control to the basic closed-loop equidistant control, recent simulator studies, carried out at GEC (Stafford, U.K.) with reference to the new Cross-Channel link, have led to the following conclusions:

(*a*) In normal steady state invertor operation, whether balanced or not, it is essential that at least half the valves in a group are on closed-loop control, otherwise harmonic instability occurs (except with a very strong a.c. system, which is irrelevant for design purposes).
(*b*) The addition of a predictive γ-control which is not normally in operation, but standing by at a few degrees less than normal, reduces the probability of commutation failure due to very small disturbances from the a.c. system.

(c) However some magnification of the original disturbance is inherent in all predictive γ-controls of whatever type.
(d) The net effect for moderate a.c. system disturbances (remote faults) is to cause so much magnification that the distortion is beyond the capabilities of predictors, and commutation failure then follows in many cases where it would not have occurred without the predictive addition.
(e) Following large disturbances, commutation failure occurs initially regardless of the type of control.
(f) During a sustained single phase fault at medium distance, after the initial shock an invertor may settle to some form of steady unbalance operation, giving finite power, provided the closed-loop control is given suitable characteristics. The addition of predictive γ-control generally gives unstable operation in this mode; hence the well known argument that predictive (individual-phase) control gives more power in this mode is a fallacy, again except for the case of infinite busbars, which is not of practical interest.
(g) The use of inverse-cosine linearisation gives a small advantage on infinite busbars, since a slightly higher loop gain can be used. However on a finite system impedance it has a detrimental effect, since for operation near to 0° or 180° the control loop causes large changes of α for a small change of the feedback quantity (I_D, γ, etc.); this produces large changes of reactive power and a.c. voltage, which tends to de-stabilise the system. A simple linear integral loop is a good compromise.

4.5 Analogue and digital controls

The effectiveness and reliability of the convertor controls are functions of the speed and accuracy with which the optimum instants of firing are generated.

Analogue controls carry out 'measurements' and 'calculations' (particularly the basic integration required in the control loops) right up to the actual moment of firing and, moreover, they do so on many channels in parallel (e.g. I_d, γ, α, V_d and I_d/γ loops). The accuracy of modern analogue controls is also perfectly adequate.

On the other hand convertor operation provides suitable information for a direct digital control in the form of periodic discontinuities of voltage and current, voltage crossings, etc. Such information can be detected and processed digitally to achieve more flexible control.[7,8] However, digital controls are slower than analogue controls because of the calculation time and hence the sampling periods are finite. The best of digital controls do not yet appear to provide the performance of practical analogue controls. Recent developments in this respect are described in Chapter 10.

Finally the general engineering practice is to use separate equipment for control, protection and monitoring and at present there is some resistance to common them into a single piece of equipment. It is now up to the designers to prove that the direct digital solution offers a similar (or higher) level of reliability.

B — D.C. SYSTEM CONTROL

4.6 Basic philosophy

From an operational point of view, the use of constant current control provides greater safeguard against disturbances. Therefore, if there is a choice and the economics are right, constant current should be used as the basic control philosophy. This, however, is not possible with conventional a.c. power systems which, being normally multiended, require reasonably constant voltages at points of common coupling.

The use of current conversion as the basis for rectification and inversion has already been justified in Section 2.2. Moreover, present d.c. schemes consist of point-to-point system interconnections and constancy of d.c. voltage levels is not a primary consideration because these are not directly available to consumers. There are, however, other considerations influencing the control philosophy, among them featuring prominently the overvoltages resulting from open circuiting and load rejections and the high resistive losses resulting from constant current transmission at low power levels.

A hybrid voltage/current philosophy is possible with d.c. transmission schemes to suit the needs of the particular operating conditions. This is achieved by adjusting the d.c. voltage levels on both sides of the link, by means of on-load tap-changer control on the steady state, and by thyristor control following large or small changes of operating conditions at either end of the link. The d.c. current is only limited by the small resistance of the transmission line and is therefore very sensitive to such variations.

It will be shown in the following sections that the provision of current controllers at both ends, combined with transformer on-load tap changing, offers a perfectly satisfactory solution to this problem; thus the use of current control is universally accepted in h.v.d.c. transmission.

4.7 Characteristics and direction of d.c. power flow

Most d.c. schemes in existence are provided with bidirectional power flow capability. This property is inherent in the case of a.c. transmission, where the direction of power flow is determined by the sign of the phase-angle difference of the voltages at the two ends of the line; the power flow direction is in fact independent of the actual voltage magnitudes.

On the other hand, in d.c. transmission the power flow direction is dictated by the relative voltage magnitudes at the convertor terminals and the absolute or relative phase of the a.c. voltages play no part in the process. However this condition can be altered by exercising a type of firing angle control which can make the power flow direction independent from the terminal a.c. voltage magnitudes and behave, instead, like the a.c. counterpart.

The basic characteristic of a convertor from full rectification to full inversion is illustrated in Fig. 4.8. The convertor is assumed to be provided with constant current and constant extinction angle controls, which have already been discussed in Section 4.3.

The level of the natural voltage characteristic can be adjusted by the transformer on-load tap changer. This part of the characteristic takes place when $\alpha = 0$ (i.e. with diode operation), i.e. the convertor has no controllability in this region and the d.c. voltage (from eqn. (2.12) with $\alpha = 0$) is

$$V_d = V_{c0} - \frac{3 X_c}{\pi} I_d \qquad (4.2)$$

For delay angles larger than $\alpha = 0$ the convertor exercises constant current control, i.e. to maintain a current reference (I_{ds}), which has a practically vertical characteristic. This region is governed by eqn. (2.12) and is limited by the need to maintain a certain minimum extinction angle required for safe commutation, as explained in Section 2.7. When this limit is reached, again the convertor loses controllability and is governed by eqn. (2.19).

Let us now consider the specific control parameter which implements the direction of power flow. We have already explained that each convertor station is normally provided with current and extinction angle control facilities. The assignment of current control, either to the rectifier or to the invertor station, is made on consideration of the investment cost for reactive power compensation, the availability of reactive power, the minimization of the losses and the total running cost. Normally the total reactive power compensation is least, and the utilization of the line best, if the rectifier is assigned the current control task while the invertor operates on minimum extinction angle control.

This combination is achieved in a two-terminal d.c. link by providing the power sending station with a slightly higher current setting than the power receiving station. The difference between the two settings is termed the current margin (I_{dm}). Its effect can be better understood with reference to Fig. 4.9(a), where the operating current is set by the constant current control at the rectifier end. The invertor end current controller then detects an operating current which is greater than its setting and tries to reduce it by raising its own voltage, until it hits the ceiling determined by the minimum extinction angle control at Point A. This is the normal steady state operating point, which presumes a higher natural voltage characteristic at the rectifier end, a condition which may require on-load tap-change action at the convertor transformer.

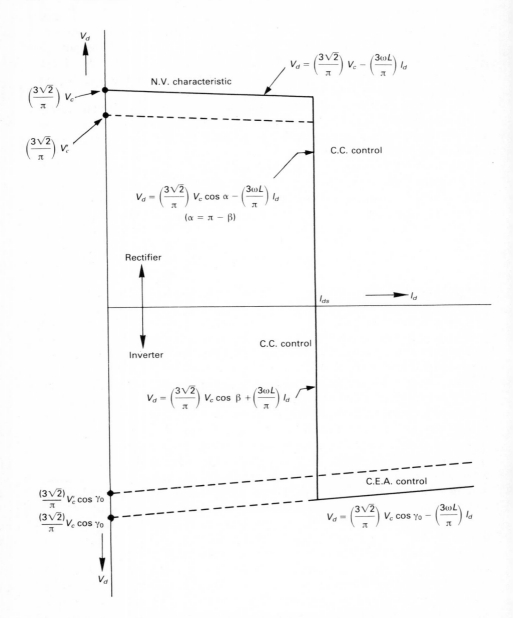

Fig. 4.8 Complete control of a convertor; from inversion to rectification
N.V. = natural voltage
C.C. = constant current
C.E.A. = constant extinction angle

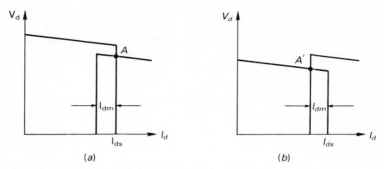

Fig. 4.9 Steady state characteristics and operating point
(a) Under rectifier current control
(b) Under invertor current control

The no-load and direct voltage regulation along the transmission link are illustrated by the continuous line in the diagram of Fig. 4.10, with power flowing from left to right and with the operating current maintained by the rectifier controller.

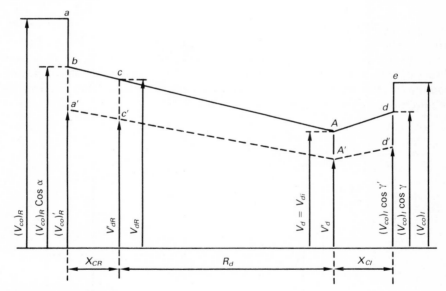

Fig. 4.10 D.C. voltage profile

Starting from the rectifier end, point a represents the maximum average direct voltage, $(V_{c0})_R$ with the rectifier unloaded. This is reduced by firing angle control to $(V_{c0})_R \cos \alpha$ (point b). The slope of lines bcA and $-dA$ is determined by the operating direct current I_d, and their horizontal projections relate to the commutation reactances X_{CR} (for bc), X_{CI} (for dA) and line resistance R_d (for cA). Therefore point c represents the output voltage at the rectifier station and point A the voltage at the remote end of the line. This point is also reached from the invertor end open circuit voltage $(V_{c0})_I$ reduced by the extinction angle $(V_{c0})_I \cos \gamma$ and by the commutation reactance. It should be obvious that point A represents the same operating condition as indicated by the crossing point in the characteristics of Fig. 4.9(a).

Let us now assume that there has been a substantial a.c. voltage reduction at the rectifier end, such that the d.c. voltage ceiling (the natural voltage) of the rectifier becomes lower than that of the invertor. In the absence of a current controller at the invertor, the voltage across the line is reversed and the current reduces to zero (current through the valves cannot reverse). However, an invertor current controller will prevent a current reduction below its setting by advancing its firing (i.e. reducing α and hence invertor d.c. voltage), thus changing from extinction angle to constant current control. A new operating point A' (Fig. 4.9(b)) results at a current reduced by the current margin. In Figure 4.10 this condition is represented by the dotted lines and clearly shows that power flow will continue in the same direction, in spite of the lower a.c. voltage at the sending end.

4.7.1 Reversal of power flow

The last section has explained that the direction of power flow is decided purely by control action. It can be expected, therefore, that a change in power direction can not happen naturally as a result of a change in the operating conditions, but rather following a control order resulting from the overall requirements of the power system.

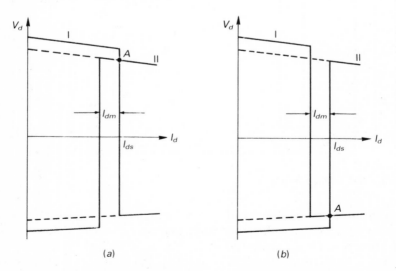

Fig. 4.11 (a) Operating point with power flowing from Station I to Station II
(b) Operating point after power reversal

The characteristics of Fig. 4.9(a) can be extended below the voltage zero-line so that the rectifier and invertor ends exchange their function. The result of this action, illustrated in Fig. 4.11(a), shows that the two characteristics do not meet again. Station 1 increases the delay angle well into the inverting region and hits the limiting extinction angle voltage (γ_0). Station 2 advances its firing into the rectifying region and finally hits the rectifier voltage ceiling ($\alpha = 0$). In the process, the line current reduces to zero and the complete system is blocked.

Since, as indicated in the previous section, the rectifying station requires the higher current setting, it is thus necessary to subtract the current margin from the reference value of Station 1. This results in the characteristics of Fig. 4.11(b), which display one operating point of different voltage polarity and with the roles of the two stations interchanged. Thus the direction of power flow has been reversed without altering the direction of current flow, which of course is fixed by the convertor valves.

4.7.2 Modifications to the basic characteristics[9]

During a.c. system faults at the receiving end there is a big risk of commutation failure and, if the fault is electrically close, the invertor may not be capable of recovering by itself. In such cases it is important to reduce the stress on the invertor valves, and this is achieved by providing a low-voltage-dependent current-limit to the rectifier control characteristic. The modified characteristics illustrated in Fig. 4.12 consists of a branch CD' at the rectifier and another EF' at the invertor.

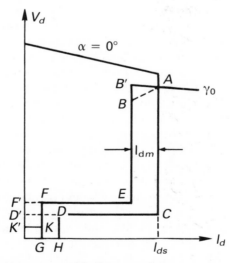

Fig. 4.12 Rectifier and invertor modified characteristics

In the literature the added branches CD' and EF' are normally given finite slopes on the basis that the finite slope is expected to provide 'smooth' control; in fact this argument is not valid, since the slope constitutes a strong negative resistance which produces the opposite effect. A plain horizontal line, as shown in Fig. 4.12, is better since the system will then operate stably down to a lower d.c. voltage during faults in the rectifier a.c. system; it will also recover faster from a total collapse of an invertor because the a.c. line/cable will charge more rapidly.

Also an unwanted voltage reversal of the invertor is generally prevented by limiting α to some minimal value larger than 90° (shown by KK' of the characteristic). Branches DH and FG represent maximum current limits at low voltage. A minimum current limit (not shown in the figure) is also normally provided to prevent operation with discontinuous current.

Another problem with the basic characteristics occurs when the rectifier voltage ceiling gets very close to that of the invertor, such that the characteristics intersect in the region between I_{ds} and $I_{ds} - I_{dm}$. In this region both current controllers are out of action. In practice the current will not settle at an intermediate point; instead, the invertor will be periodically entering the current controlling mode. A proven countermeasure against this type of oscillation is a small shape alteration of the invertor characteristic, as shown in Fig. 4.12. The limiting $\gamma = \gamma_0$ characteristic cannot be changed without sacrificing heavily the reactive power; it is, however, sufficient to break the characteristic to a positive resistance slope in the transfer region from γ-control to current control of the invertor (AB in Fig. 4.12 instead of AB').

4.7.3 Tap changer control

Referring to Fig. 4.9(a), the task of the tap changer on the rectifier side is to place the voltage ceiling (i.e. the $\alpha = 0$ characteristic) relative to the operating point (A), so as to minimise the reactive power consumption subject to a minimum α limit. This limit is used to ensure that all the anodes (in the case of the mercury-arc valve) or the series thyristors have a sufficient positive voltage across them to avoid the risk of misfiring.

On the invertor side the reactive power is minimised at the $\gamma = \gamma_0$ limit, which is the normal operating condition. It is therefore possible to use its tap changer to keep the direct voltage within any desired limits.

When the current control is transferred to the invertor (as in Fig. 4.9(b)), the rectifier tap changer tries to raise the voltage ceiling and the tap changer action on the invertor side must stop, since the latter no longer determines the voltage level. Otherwise the tap changer would try and raise the invertor side a.c. voltage and thus reduce the power factor.

It is interesting to note that the d.c. link, within its normal power rating, operates as a constant voltage circuit, i.e. the voltage is at its highest value under no-load conditions and power is raised by increasing the current settings. However, in order to avoid overloads, when the current limit is reached, the transmission link acts as a constant current system, i.e. a power change can then only be achieved by means of a voltage change and, because

Control of h.v.d.c. convertors and systems 95

of the voltage ceilings, this is only possible to a limited extent by means of tap changer action.

4.7.4 Different control levels

For reasons of reliability the controls of an h.v.d.c. link are generally divided into four different levels,[10] i.e.

(a) Bridge controls — To control the firing instants of the valves within a bridge and to define the γ_0 and α_{min} limits.
(b) Pole controls — To coordinate the bridges in a pole to provide the ordered current, with minimum harmonic generation.
(c) Master controls — To provide coordinated current orders to all the poles.
(d) Overall controls — To provide the current orders to the master controls in response to required functions such as power transfer control, system frequency control, system damping or combinations of them. This subject is further discussed in Chapter 5.

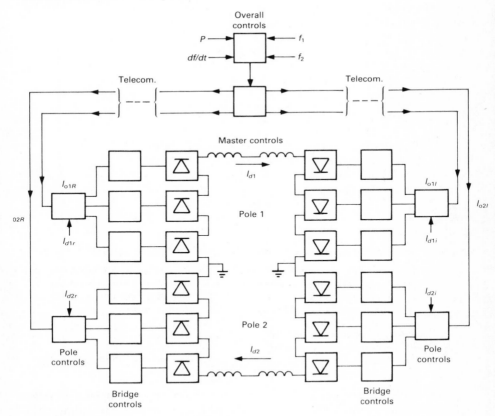

Fig. 4.13 Block diagram of a control scheme

96   Control of h.v.d.c. convertors and systems

A simplified block diagram of a multibridge control system is illustrated in Fig. 4.13. The bridge controls, and in particular the valve firing circuits, contain most of the components involved, and are therefore kept independently for each bridge unit (or convertor group in modern 12-pulse schemes) for reliability reasons.

4.7.5 Power flow control

As the primary object of h.v.d.c. transmission, power control should be the main consideration. However for simplicity the early schemes still used the current control loop as a basis. In such case each station is provided with a dividing circuit, consisting of a power calculator and a high gain operational amplifier, which receives the power order, voltage and current signals and produces the current order. Other components of the station power control unit,[9] illustrated in Fig. 4.14, include the telecommunication equipment (TC), the constant power order setting unit (OS), an emergency power controller (EPC) and a current limiter. A control error (CE) is formed from comparison of the calculated current order (minus the current margin in the case of the invertor station) and the measured current.

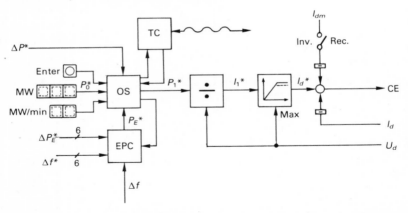

Fig. 4.14 *Example of power flow control with additional emergency control for frequency*

TC	Telecommunication equipment	Δf^*	Thresholds for frequency deviations
EPC	Emergency power controller	Δf	AC network frequency deviation
P_0^*	Manually set power order	I_1^*	Calculated current order
ΔP_P^*	Additional power order by higher level controller	CE	Control error
		Inv	Invertor
ΔP_E^*	Step in emergency power order	REC	Rectifier
P_E^*	Emergency power order		
OS	Order setting unit		

(Asterisks denote reference values)
(© 1978 CIGRE)

In some early schemes each pole at both stations was equipped with power flow controls as shown in Fig. 4.14. The power order and corresponding rate of change order could then be set by the operator at the main station, or they could be transmitted there from a dispatch centre. The task of the order setting unit of such scheme is to implement power stepping, and synchronisation of order setting at the two ends of the link, by telecommunications.

In most cases there is only one master controller (at one of the stations) which sends a current order to the pole controls of the two ends of the link.

In recent schemes the power is monitored by multiplying voltage and current (summed from both poles) and fed back directly to the controller. Like in previous control methods, to prevent unacceptable current orders (e.g. during start-up), limits are normally built-in.

The d.c. power setting philosophy and its influence on the overall power system control are discussed on Chapter 5.

4.7.6 Telecommunication requirements[11]

The telecommunication means used in h.v.d.c. schemes include microwave radio, carrier on the power conductors private wire, rented wire and the use of the public telephone system.

All of them are liable to interference and even occasional failures. Carrier systems, using the h.v.d.c. line or cable, are affected by a continuous source of interference, i.e. the convertors, which is difficult to filter because of the high impedance of the smoothing reactors; the use of the lowest possible bandwidth is thus recommended to reduce the effective noise.

To avoid noise the analogue signals are converted to digital form and transmitted with an error checking signal.

Digital signals are normally sent in regular blocks of binary bits, at a block rate which depends on the telecommunication medium and on the desired rate of response. The main information includes the current order and some logical signals like the power flow direction and fast shut-down orders.

The need for other fast signals such as the a.c. system damping will be discussed in the next chapter. A number of relatively slow signals such as the interchange of manually-set power orders or supervisory signals can be sent by sub-multiplexing.

Without special arrangements to coordinate the operation at both ends of the link, a reduction of rectifier (or an increase of invertor) current order by more than the current margin, relative to the invertor (or rectifier) current order, can cause a complete d.c. voltage shutdown. This problem is normally avoided by the following procedure:

(*a*) When the master controller operates from the sending (i.e. rectifying) end, an 'increase' in current order is implemented immediately at the local end, and as soon as possible (only subject to the unavoidable telecommunication delay) at the remote end. On the other hand a 'decrease' in current order updates the remote current order first, while the local current order change waits for a 'true' check-back signal from the remote end, via a return telecommunication channel indicating the receipt of an error-free signal.

(*b*) When the master controller is housed at the invertor end, the above procedure still applies but with the words 'increase' and 'decrease' interchanged.

(c) If an error is detected in either telecommunication receiver then the local current order is not updated; the return check-back is sent as a zero.

The functions of (a)–(c) are obtained by the telecommunication coupling logic, which requires two registers respectively for 'present' and 'last' current order, with a comparator and a few simple gates. It is inserted between the source of master current order and the telecommunication, and also supplies the output to the local pole controls. Its effect is to ensure that the sequence of current order updating is never such as to decrease the effective current margin, even in the presence of errors in the telecommunication systems, and to freeze both current orders for a detected error in either telecommunication channel.

4.8 References

1. TARNAWECKY, M. Z. (1971): 'H.v.d.c. transmission control schemes', *Manitoba Power Conference EHV-DC*, Winnipeg, pp. 699–741.
2. HINGORANI, N. G. and CHADWICK, P. (1968): 'A new constant extinction angle control for a.c./d.c./a.c. static convertors', *Trans. IEEE*, Vol. PAS-87, No. 3, pp. 866–872.
3. AINSWORTH, J. D. (1968): 'The phase-locked oscillator–a new control system for controlled static convertors', *Trans. IEEE*, Vol. PAS-87, No. 3. pp. 859–865.
4. UHLMANN, E. (1975): *Power Transmission by Direct Current*, p. 147, Springer-Verlag, Berlin/Heidelberg.
5. EKSTROM, A. and LISS, G. (1970): 'A refined h.v.d.c. control system', *Trans. IEEE*, Vol. PAS-89, No. 5/6.
6. RUMPF, E. and RANADE, S. (1972): 'Comparison of suitable control systems for h.v.d.c. stations connected to weak a.c. systems. Part I: New control systems. Part II: Operational behaviour of the h.v.d.c. transmission', *Trans. IEEE*, Vol. PAS-91, pp. 549–564.
7. ARRILLAGA, J., GALANOS, G., and POWNER, E. T. (1970): 'Direct digital control of h.v.d.c. convertors', *Trans. IEEE*, Vol. PAS-89, No. 8, pp. 2056–2065.
8. REEVE, J. and SEVCENCO, J. A. (1972): 'An automatic control scheme for h.v.d.c. transmission using digital techniques. Part I: Principles of operation', *Trans. IEEE*, Vol. PAS-91, pp. 2319–2324.
9. JOTTEN, R., BOWLES, J. P., LISS, G., MARTIN, C. J. B. and RUMPF, E. (1978): 'Control of h.v.d.c. systems — The state of the art', *CIGRE Paper 14–10*, Paris.
10. BOWLES, J. P. (1975): 'Control systems for h.v.d.c. transmission', *Report to CEA-HVDC Subsection*, Edmonton.
11. AINSWORTH, J. D. (1981): 'Telecommunication for h.v.d.c.', *IEE Conference Publication 205 on Thyristor and Variable Static Equipment for A.C. and D.C. Transmission*, London, pp. 190–193.

Chapter 5
Interaction between a.c. and d.c. systems

5.1 Introduction and definitions

A d.c. link can be operated according to the basic control modes described in the last chapter and thus remain passive to any special needs of the interconnected a.c. systems. Alternatively the link can be provided with more dynamic controls, capable of responding to any deviation from the normal operating condition in the a.c. or d.c. systems.

The exclusive use of the basic controls often gives rise to unwanted interaction between the a.c. and d.c. systems, which is manifested in a variety of voltage, harmonic and power instabilities. When full advantage is taken of the fast and adaptable convertor controllability a more useful interaction can be achieved, which manifests itself in stable a.c. and d.c. system operation.

The degree of interaction obviously depends on the strengths of the a.c. and d.c. systems and therefore the concept of relative system strengths must first be clarified.

With reference to Fig. 5.1, the various impedances connected to the convertor station can be combined into an equivalent Thevenin source impedance Z_{st}, for the purpose of defining the strength of the a.c. system.

Z_{st} is the equivalent of Z_s (source), X_{sc} (synchronous compensators), X_c (filters and static compensation) and Z_l (local load) in parallel. The equivalent Thevenin source voltage E_{st} results from the vectorial addition of $V_c/\sqrt{3}$ (line to neutral) and $I.Z_{st}$ as shown in Fig. 5.2.

As the d.c. link forms an integral part of the power system, it is convenient to express the convertor equations in the same per unit system as the a.c. network.

A convenient factor in the per unit notation is the convertor rating factor r, defined as the ratio of convertor MVA to d.c. power (P_d), i.e.

$$r = \sqrt{3}V_c I/(V_d I_d) \tag{5.1}$$

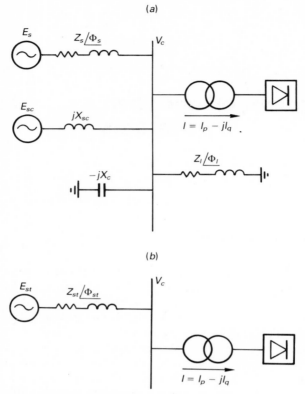

Fig. 5.1 A.C. system representation at the convertor busbar
(a) Individual impedances
(b) Thevenin equivalent

and substituting eqn. (2.14)

$$r = (3\sqrt{2}/\pi)V_c/V_d. \qquad (5.2)$$

It is standard practice to refer the parameters of a power plant component to its rated power and voltage. In the case of the convertor bridge we will use as reference base parameters the convertor transformer MVA and nominal voltage. The base impedance is therefore given by

$$Z_{BASE} = V_c/(\sqrt{3}I) = V_c^2/MVA_c. \qquad (5.3)$$

If MVA_F is the short circuit level of the a.c. system at the convertor terminal, the system impedance Z_{st} can be expressed as

$$Z_{st} = V_c^2/MVA_F \qquad (5.4)$$

and in per unit

$$z_{st} = Z_{st}/Z_{BASE} = (V_c^2/MVA_F)(MVA_c/V_c^2) = MVA_c/MVA_F. \qquad (5.5)$$

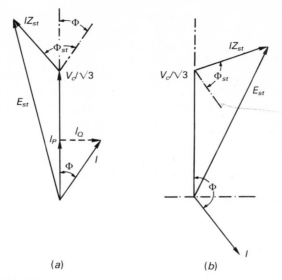

Fig. 5.2 *Voltage regulation*
(a) Rectifier end
(b) Invertor end

Using the convertor rating factor r, the per unit system impedance can be expressed as

$$z_{st} = rP_d/\text{MVA}_F = r/\text{SCR}, \tag{5.6}$$

where the SCR factor is normally referred to as the Short Circuit Ratio, i.e. the ratio of the short circuit MVA at the a.c. busbar to the nominal d.c. power of the convertor station connected to the same busbar.

It must be kept in mind, however, that the fault MVA, or the SCR of an a.c. system, normally define the transient impedance (for use in fundamental frequency overvoltage assessment) or the subtransient impedance (for use in harmonic studies).

However, the impedance of most a.c. systems (before adding capacitors or filters) is approximately equivalent to a fixed inductance only at fundamental frequency, at second harmonic and perhaps at third and fourth harmonics with increasing errors. Although the SCR does not define impedance at higher harmonics, it is used as a reasonable practical criterion of convertor behaviour, since the combination of a.c. system and harmonic filter will have a low principal natural frequency (particularly in the presence of conventional tuned filters); this natural frequency is normally the determining factor in the development of overvoltages and harmonic instabilities.

The Short Circuit Ratio of h.v.d.c. schemes at the receiving end has been gradually reducing, i.e. modern invertors feed power into relatively weak systems, which are therefore more susceptible to increased overvoltages and invertor maloperation. It is quite common now for the SCR to be in the range 2 to 4 and it is conventional to regard stability and voltage control as problems associated with h.v.d.c. links operating into weak a.c. networks, principally at the invertor, where the SCR is less than 3.

However, while such practice is acceptable for the operation of the d.c. link, fundamental and harmonic frequency voltage problems may still occur on the a.c. network at higher SCRs, and special steps must be taken in such cases to maintain an acceptable quality of supply to consumers.

5.2 Voltage interaction

The a.c. terminal voltages at the convertor stations depend on the active and reactive power characteristics of the convertor. To minimise voltage variations it is essential to control the reactive power supply to match the convertor reactive power demand.

It has already been explained in Chapter 2 that an h.v.d.c. transmission link consumes reactive power at both ends, which can be typically 60% of the power transmitted at full load. Moreover, during transients the reactive power demand may vary widely, the duration of such variation depending to a large extent on the characteristics of the d.c. link control system.

With the combination of rectifier constant current and invertor constant extinction angle controls, both backed by transformer on-load tap-changers, the reactive versus active power variation is non-linear as shown in Fig. 2.13. The minimum values of firing delay, extinction angle and commutation reactance, however are determined by other considerations such as the need to reduce the risk of valve maloperation and to limit the valve stresses.

If the presence of local generators can be guaranteed (e.g. close to a rectifier) it is always more economical to supply most reactive power from these, with minimum size filters to reduce harmonics.

For other cases it may be necessary to provide full reactive power compensation of the convertor, sometimes with extra compensation for a.c. system loads also. A large proportion of the reactive power is supplied by the shunt harmonic filters, and additional compensation has so far been provided in the form of shunt capacitor banks or synchronous compensators. The filters and capacitors are normally connected to the high voltage terminal, whereas the synchronous compensators are connected through a transformer tertiary winding. Other permutations have also been used, e.g. filters on a tertiary winding and synchronous compensators on separate transformers.

The size of the individual filters is the result of a compromise between economy (which demands the larger size) and the ability of the a.c. system to accept the step changes in system voltage caused by filter switchings. Switching is often needed to control reactive power at d.c. loads lower than the nominal. There is a further difficulty in discrete filter switching which relates to the non-linear relationship between the increase of reactive power and harmonic current with load.

If their extra cost can be justified, the use of dynamic compensation, in the form of either synchronous compensators or controllable static compensators, can reduce or eliminate the step switching of filter branches. These are discussed in more detail in Section 5.2.2.

5.2.1 Dynamic voltage regulation

During h.v.d.c. link disturbances the voltage control requirements depend on the nature and location of the disturbance. The reactive power consumption, although possibly higher initially, is partially or totally eliminated following the disturbance, with the result of considerable dynamic overvoltage regulation.

With reference to the voltage diagram in Fig. 5.2, let us consider the effect of total load rejection (i.e. $I = 0$) at the rectifier and at the invertor ends. The maximum voltage regulation will occur if the disturbance takes place when the phase angle of the convertor (ϕ) is equal to the phase angle of the equivalent a.c. system impedance (ϕ_{st}), and is calculated as follows:[1]

$$E_{st} = V + I \cdot Z_{st}. \tag{5.7}$$

Dividing throughout by the base voltage

$$E_{st}/V_c = V/V_c + Z_{st}I/V_c \tag{5.8}$$

and substituting expressions (5.3) and (5.6)

$$e_{st}/\sqrt{3} = (v/\sqrt{3}) + (I/V_c)(r/\text{SCR})(V_c^2/\text{MVA}_c) \tag{5.9}$$

or

$$e_{st} = v + r/\text{SCR}. \tag{5.10}$$

With an SCR of 2.5 and under rated conditions

$$e_{st} = 1 + 0.4r. \tag{5.11}$$

In order to calculate r let us refer to the d.c. voltage equation (2.12), i.e.

$$V_d = V_{c0} \cos \alpha - (3X_c/\pi)I_d, \tag{5.12}$$

which can be more conveniently written as

$$\frac{V_d}{V_{c0}} = \cos \alpha - (3X_cI_d)/(\pi V_{c0}). \tag{5.13}$$

Substituting $V_{c0} = (3\sqrt{2}/\pi)V_c$ and eqn. (5.2) yields

$$\frac{1}{r} = \cos \alpha - X_cI_d/(\sqrt{2}V_c). \tag{5.14}$$

The commutation reactance in per unit is

$$x_c = X_c/[V_c/(\sqrt{3}I)] \qquad (5.15)$$

and since $I = (\sqrt{6}/\pi)I_d$, eqn. (5.15) can be changed into

$$X_c I_d/(\sqrt{2}V_c) = (\pi/6)x_c, \qquad (5.16)$$

which substituted in eqn. (5.14) finally produces

$$\frac{1}{r} = \cos\alpha - \frac{\pi}{6}x_c. \qquad (5.17)$$

Thus for a system having a commutation reactance of 0.2 p.u. (to the base of transformer MVA), and a rated firing angle α (or γ) of 15°, the value of r in eqn. (5.17) is approximately 1·15, which substituted in eqn. (5.11) results in

$$e_{st} = 1\cdot 46.$$

Under these particular severe conditions a regulation of 46% can therefore be expected.

In order that $\phi = \phi_{st}$ (at a rectifier) under rated conditions, either the phase angle of the system impedance needs to be of the order of 30°, an unusually low value, or the rectifier should operate at an unusually high firing angle. Typical impedance angles of the order of 70–85° are found in practice; the more resistive the a.c. system appears at the rectifier end, the higher regulation overvoltages may be anticipated.

At the invertor, however, it can be seen (Fig. 5.2(b)) that under normal operating conditions ϕ_{st} is considerably less than ϕ. Therefore the more reactive the a.c. system appears, the higher is the regulation overvoltage. With an impedance of 0·4 p.u., a maximum value of $e_{st} = 1\cdot 29$ p.u. can be anticipated, thus causing a 29% regulation, following full load rejection; however, the expected impedance angle in practice is of the order of 70°.

The regulation (dynamic) overvoltages are therefore more significant at the rectifier end. At both terminals the effective impedance angle is as important in determining the overvoltages as is the magnitude of the impedance. For links from hydro sources, the increase of frequency following load rejection will produce even higher dynamic overvoltages. This is an unacceptable situation for local consumers and must be allowed for in the insulation coordination of the convertor station.

In practice, transformers start to saturate at typically 1·2 to 1·25 p.u. a.c. voltage and the fundamental frequency overvoltage will therefore be a little lower, with some distortion.

Single line to ground faults are also a source of dynamic overvoltages on the other phases or pole. As a result of an a.c. phase to earth fault, the mutual coupling between phases causes a voltage increase in the other phases; for a network effectively earthed the overvoltage is limited to a peak value of $\sqrt{2}(0.8)$ of the line-to-line rms.

On the other hand, following a voltage drop in the a.c. network, the initial effect is a fall in power. The power controller of the d.c. link then increases the current reference to try and restore the ordered power; the extra current increases the reactive demand and tends to reduce the a.c. system voltage further. With very weak a.c. systems this could lead to voltage collapse; however, power controllers always have limits built-in to avoid excessive action.

By far the most important case is that of a nearby three-phase short-circuit, assuming that the convertors are blocked permanently during the fault, with all the capacitors left on. This condition produces full magnetizing inrush current on all transformers after fault removal, which results in substantial fundamental and harmonic overvoltages. Such overvoltages constitute in practice the determining condition for most valve, surge arresters and insulation voltage ratings.

5.2.2 Dynamic compensation

Dynamic compensation equipment is used to reduce the dynamic voltage regulation, to help in the recovery of the a.c. system from faults, and to reduce the disturbances resulting from d.c. load variation or from the switching of filter banks.

Ideally to meet such a comprehensive range of duties, the compensation will generally require to have both reactive power absorption and generation capabilities. In each particular application the system Short Circuit Ratio and the stability of the d.c. link and compensator controls must be considered, when trying to decide the type and dynamic range to be used.

A preliminary comparison of the technical characteristics of reactive power compensators for use with h.v.d.c. transmission, with extensive bibliography on the subject, has been presented at CIGRE.[2]

The main types of dynamic compensation already used, or under consideration for h.v.d.c. schemes, are: synchronous compensator, a.c. self-saturated reactors, thyristor-controlled reactors and thyristor-switched capacitors.

The slow response of the synchronous compensators may be a problem particularly in the absence of local generation. However, the synchronous compensator reduces the sensitivity to transients by increasing the SCR. It also increases the resonance frequency of the system, since it reduces the need for extra shunt capacitance.

It is often claimed that the last three alternatives (i.e. the static ones) improve the voltage stability of the a.c. network and thus help h.v.d.c. control stability and speed of response. In practice the so-called 'fast' static compensators rarely improve voltage stability, except that they act as limiters if the voltage tries to rise too much. In steady state the practicable thyristor-controlled reactors appear to cause some destabilisation, though this can be made acceptable; saturated reactors appear to be better in this respect.

The other main problem of the fast static compensators appears when the system e.m.f. falls even slightly; in such case, both, the thyristor-controlled and the saturated reactors, tend to go to a zero-current condition, and stability then depends only on the bare a.c. system plus all capacitors. Normally the latter account for 0·6 p.u. per convertor plus an extra 0·4 p.u. for load compensation and under these conditions the net SCR could be very low (e.g. 1·5), making stability difficult.

It would be fair to say at this stage that more experience and deeper studies are needed in this area.

5.3 Harmonic instabilities

Harmonic instability at a convertor terminal can be defined as the generation and/or magnification of non-characteristic frequencies by a d.c. system containing, initially, no unbalance or asymmetry.

The problem is best illustrated with reference to typical instabilities encountered in various existing schemes; these are described in Sections 5.3.1–5.3.3. Finally a general method suitable for predicting possible instabilities is discussed in Section 5.3.4.

5.3.1 Instabilities caused by individual firing control

This section reproduces the basic analysis and conclusions contained in a 'classic' paper by Ainsworth[3]. Using simple analysis, based on one convertor and one harmonic, and assuming linear behaviour for the small levels of harmonic voltage distortion involved, the paper demonstrated the existence of a problem which changed dramatically the conventional control philosophy.

In the circuit of Fig. 5.3 a spurious harmonic e.m.f. of magnitude E_n and harmonic order n (of positive or negative sequence) is added to the voltage source E_{st}.

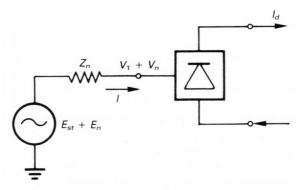

Fig. 5.3 Equivalent circuit per phase for a.c. system of finite impedance

The a.c. system impedance is assumed to be Z_n at the harmonic frequency of order n and zero at other frequencies. The d.c. reactor inductance is assumed infinite; i.e. the d.c. line current is constant.

The harmonic voltage V_n on the convertor a.c. busbars then depends on impedance Z_n and its phase angle ϕ_{sn}, harmonic order n, firing angle α, and relative phase θ of the interfering e.m.f. The latter is random and therefore only the maximum V_n as θ varies is of importance. The harmonic magnification due to convertor operation is

$$M = V_n/E_n$$

Neglecting transformer saturation effects, M is given by the following equations (for a constant-α control system):

(a) For $n = 6p - 2$

$$M_{max} = \left[1 + x\cos(\phi_{sn} - n\alpha) + \tfrac{1}{2}x^2 - x\sqrt{\left\{1 + x\cos(\phi_{sn} - n\alpha) + \frac{x^2}{4}\right\}}\right]^{-} \quad (5.18)$$

where

$$x = Z_n I_d \sqrt{6}/(V_1 \pi). \quad (5.19)$$

(b) For $n = 6p + 2$

$$M_{max} = \left\{1 + x\cos(\phi_{sn} - n\dot\alpha) + \frac{x^2}{4}\right\}^{-1/2}. \quad (5.20)$$

In both cases the voltage magnitude per phase is the same.

For $n = 3p$, both sequences (orders $+n$ and $-n$) are obtained in busbar voltage and current. Phase voltages are then unequal, but the maximum M for the highest of the three phase voltages is the same as in the first equation.

For high Z_n, high magnifications can occur. In the worst cases, Z_n greater than some critical value gives infinite magnification, i.e. instability. In these cases, it is obvious that the instability can be initiated from any asymmetry, however small, whether originating in the a.c. system, as postulated, or elsewhere, even with a perfectly adjusted control system. In practice V_n will not rise indefinitely, but will be limited at a considerable value by nonlinearity, which is ignored in this analysis. An example of harmonic instability obtained in a physical model is illustrated in Fig. 5.4.

In an h.v.d.c. convertor, although shunt harmonic filters are normally provided to reduce total a.c. system impedance to low values at 'normal' harmonic orders (5, 7, 11, 13, etc.), it is not economical to provide filters for the intermediate 'abnormal' harmonics, and the total Z_n may be high for these. The normal harmonics may in any case be readily shown to produce the same change of firing pulse time for every valve, but not to affect their relative spacing, and hence may be ignored.

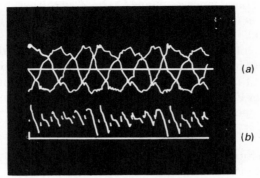

Fig. 5.4 *Model test on 6-pulse convertor with realistic a.c. system (Short Circuit Ratio = 3) and conventional harmonic filter.*
 Constant-α control system, $\alpha = 32°$; control-system-filter Q factor = 2·5
 (a) A.C.-line voltages
 (b) D.C.-bridge voltage

It is, however, clear from the equations above that if z_n (in per unit) can be guaranteed to be less than

0·5 for $n = 2, 4, 6, 8$, etc.,

1·0 for $n = 3, 9, 15$, etc.,

there is no danger of instability, though magnifications may still occur. This can in principle be achieved by the addition of extra shunt filters to the main circuit, but this is usually very expensive.

If a shunt filter is used for the normal harmonics, i.e. orders 5, 7, 11, 13, etc., partial resonance may occur between the a.c. system and the filter at the abnormal harmonics; the requirement then virtually implies that the conductance component of the a.c. system harmonic admittance must be greater than 2 p.u. or, expressed in another way, the a.c. harmonic impedance on a polar diagram must be within a circle or radius 0·25 p.u. centred on (0·25, 0).

Except for the case of a relatively small convertor connected to a large a.c. system, this is found to be a somewhat stringent requirement, and in certain typical cases the calculated maximum circle radius has been about 1–2 p.u., implying stable operation only up to 0·125–0·25 of rated current, with the worst combination of α and system impedance angle.

The main practical effects of large harmonic magnification or instability are:

(*a*) Excessive harmonic voltages and currents in the a.c. and d.c. systems. Instability due to even one system resonance, in general, produces distortion containing all harmonic orders, i.e. 2, 3, 4, etc. Usually any local overvoltage due to this is small, but interference elsewhere may be unacceptable.

(b) Operation approaching full inversion may be impossible owing to continuously repeated commutation failures.

It should be emphasised that, as shown by the equations, operation is critically dependent on α; a convertor which shows harmonic instability at one value of α due to an a.c. system resonance, may be stable with only small distortion at a value of α different by only say 20°, depending on the frequency of the resonance.

In many cases it will be desirable to take steps to prevent the magnification or instability discussed. The method used in early schemes was the addition of a control-system filter in the supply of a.c. timing voltages from the main system to the control system, and this has been discussed in Section 4.2. However this method by itself did not eliminate harmonic instabilities and extra expensive filters had to be added in some schemes.

The constant extinction angle control systems are analogous to the constant-α control system, but they, in effect, control to a constant negative time delay (the extinction angle γ), measured back from the instant corresponding to the second zero crossing (at $\alpha = 180°$) on the appropriate a.c.-line-voltage waveform.

Thus, harmonic magnification or instability can occur with constant extinction angle control systems, as for constant-α systems, if the a.c.-system harmonic impedance Z_n is high. The results can, in practice, be worse, since practical predictive systems can only operate on smooth waveforms; with badly distorted waveforms, they will make many 'mistakes', which, in the worst cases, produce continuous commutation failures and collapse of operation.

It appears to be an inescapable conclusion that purely predictive systems are not workable for constant extinction angle control of invertors on very high-impedance a.c. systems. Some improvement is, however, obtained by the use of control system filters.

5.3.2 Convertor transformer saturation effects[4]

Transformer saturation is a well-known source of harmonic current. The saturation can occur as a result of d.c. magnetisation or overexcitation. Transformer textbooks normally refer to 'odd' ordered harmonics being generated, with predominance of 3rd harmonic. They also explain that with balanced a.c. voltages the third harmonic can be absorbed in a delta winding.

The case of a convertor transformer is very special in this respect because, the convertor, under non-ideal operating conditions, can produce non-zero sequence triplen harmonics, even harmonics and direct current.

The presence of direct current on the convertor side of the transformer introduces magnetic imbalance as shown in Fig. 5.5, i.e. there is an average flux component which implies the existence of a direct excitation component.

It has been shown[5] that the harmonics generated by the transformer under d.c. magnetisation have a linear relationship with the direct current and are largely independent of the a.c. excitation.

110 Interaction between a.c. and d.c. systems

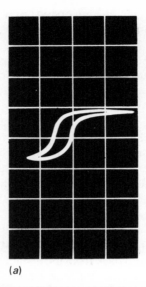

(a)

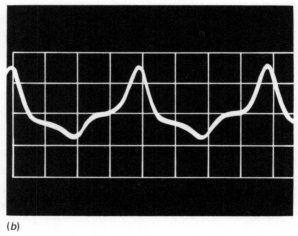

(b)

Fig. 5.5 *Magnetic imbalance*
  (a) Excitation characteristic $\phi = f(i)$
  (b) Excitation current waveform $i = f(t)$

Another problem, particularly affecting convertor transformers, is the event of double-sided saturation caused by overexcitation when blocking the convertor group. A typical convertor transformer may operate with a peak magnetic flux density of say 1·6 to 1·7 Tesla, whereas a 30% overvoltage due to voltage regulation may produce a magnetic flux density of say 1·9 to 2 Tesla, which will push the transformer well into saturation. The level of transformer saturation with respect to the applied voltage magnitude is typically as shown in

Interaction between a.c. and d.c. systems 111

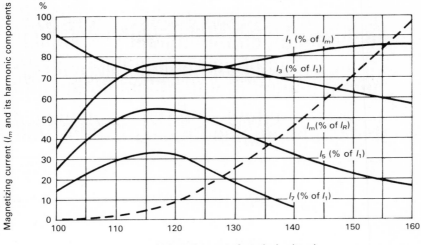

I_R Rated current
I_m Magnetizing current
I_1, I_3, I_5, I_7 Fundamental and harmonic currents

Fig. 5.6 *Harmonic components of transformer exciting current*

Fig. 5.6.[6] The magnetising current contains all odd-ordered harmonics, of which the triplen ones will be expected to be absorbed in delta windings. With 6-pulse convertor groups the convertor filters will be able to absorb the remaining harmonic orders. If, however, 12-pulse groups are used, the 5th and 7th harmonics will penetrate into the a.c. system and, in the presence of a low short circuit ratio system, the parallel a.c. system/filter arrangement may be resonant at either the 5th or 7th harmonic.

A third important problem, not exclusive of convertor transformers, is caused by inrush currents during transformer energisation and is due to the residual flux density in the core from the previous switch off. In theory, peak flux densities of up to 4·7 Tesla can occur in standard transformers (as compared with the 2 Tesla of the previous case). This will push the transformer deeply into saturation and give rise to a magnetising current of between 5 and 10 per unit, as compared with a normal magnetising current of a few per cent. In practice the peak flux density rarely exceeds 3 Tesla, since primary saturated reactance (plus system reactance) is typically 0·3 to 0·5 p.u. in convertor transformers, depending on winding arrangement.

Typical energisation currents recorded at the Henday terminal of the Nelson River scheme are illustrated in Fig. 5.7 together with the corresponding voltage distortion.[7]

Generally the largest inrush current will occur when the residual flux is maximum (either positive or negative), with the energisation instant at a

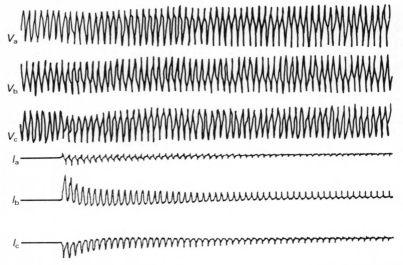

Fig. 5.7 *Voltages and inrush currents during transformer energisation (© 1980 IEEE)*

voltage zero and increasing in a direction such that its integrated area is complementary with the residual flux. This current has a high harmonic content, which is non-linear as a function of inrush pulse amplitude, and is illustrated to a first order approximation in the graphs of Fig. 5.8. Second order effects are introduced by other causes such as voltage distortion and delta transformer windings.

While the maximum value of the inrush current occurs when the circuit is closed at the instant of zero voltage, the maximum value of the temporary overvoltage is not necessarily related to this condition; it is more dependent on the natural frequencies of the system coinciding with the frequencies of the predominant harmonics, even if the inrush current peaks are below the highest values.

5.3.3 Core saturation instability[8]

A type of harmonic instability can result from a combination of a weak (high impedance) a.c. system and a resonance near to fundamental frequency on the d.c. side, due to the particular combination of d.c. reactors and d.c. cable capacitance. As with any true instability it is not caused by any inherent unbalance in the system.

The presence of a small 2nd harmonic component on the a.c. voltage, produces a fundamental frequency voltage component on the d.c. side. Due to the resonance effect this voltage causes a relatively large fundamental d.c. current component which, through the control system current feedback (whether of the individual firing or equidistant firing types), produces unequal firing pulse spacings.

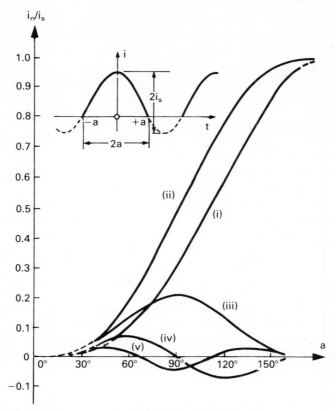

Fig. 5.8 *Inrush current components versus saturation period*
  (i) D.C. component
  (ii) Fundamental component
  (iii) Second harmonic
  (iv) Third Harmonic
  (v) Fourth harmonic

This in turn causes two effects, i.e. the generation of an additional contribution to fundamental frequency on the d.c. voltage and of a small d.c. current component on the convertor transformer. The latter tends to saturate the convertor transformer cores, generating extra magnetising current components, of which the 2nd harmonic may cause reinforcement of the original postulated 2nd harmonic voltage.

Convertor transformer saturation is the most notable feature of this type of instability, giving it its very slow response (several minutes). The high sensitivity of large transformers to d.c. is an important part of the effect.

The instability has been observed on the Kingsnorth–Willesden h.v.d.c. link, which has a d.c. side resonance near to fundamental frequency. It has never exceeded a rather small amplitude, which would possibly have remained unnoticed, except that an antiresonance exists under some system conditions exactly at 12th harmonic in the a.c. system (between the resonances of the 11th and 13th harmonic filters). The effect magnified the 12th harmonic component of magnetising current caused by the instability, and produced a circulating current, mainly between the 11th and 13th harmonic filters. This disturbed the tuning detection circuits with the result that the total filter current per arm sometimes rose to values exceeding the filter rating, and hence caused the protection to shut down the entire d.c. link.

A temporary solution to the problem is to reduce the current (power) order because the loop gain is proportional to the d.c. current level. Often the instability can be prevented at the design stage by choosing a d.c. reactor value which avoids the fundamental frequency resonance.

The cure adopted for the Kingsnorth scheme was to add an extra "flux control loop" responding to the presence of d.c. component in the magnetising current. This current is measured indirectly by monitoring the 2nd harmonic component of magnetising current per core (after rejection of load current) via a phase-sensitive demodulator. The three resulting signals per bridge are then combined to form a single (a.c.) modulating signal, which augments the common control signal per bridge. This additional modulating signal is deliberately limited to a low amplitude and has negligible effect on normal control operation.

An important consequence for future schemes is that it is not necessary to design them to avoid fundamental frequency resonance (which can frequently be difficult, e.g. due to other restrictions on d.c. reactor values and component tolerances).

5.3.4 Generalisation of the instability problem

When the d.c. current is reasonably flat, i.e. in the presence of unlimited smoothing reactance, the harmonic instability problem can be eliminated with the use of equidistant firing control. In a practical case, where the convertors are connected to a non-infinite d.c. inductance, harmonics enter the convertor controller via the d.c. current and instability can occur even with equidistant firing control, the worst cases relating to systems when a fundamental resonance exists on the d.c. side (as already indicated in the last section).

A general approach has been suggested[9] for the prediction of harmonic instabilities with non ideal convertor operation and with finite a.c. and d.c. systems impedances.

The solution involves an iterative process which starts by assuming an initial voltage supply, either exclusively of fundamental frequence or distorted, if information of the supply distortion is available. The valve firing instants can then be calculated according to the type of control implemented and, with the firing instants, the direct voltage waveform is obtained.

The direct current, derived from the d.c. voltage and the impedance of the d.c. circuit, together with the firing instants and commutating currents (which are previously derived from the known voltage sources and a.c. impedances), are used to calculate the a.c. currents. These currents are finally injected into the a.c. system/filter combination to obtain a better estimate of the convertor terminal voltages, which are then used as the starting point for another iteration.

In calculating the firing instants, it is necessary to consider, as well as the busbar voltage crossover, the effect of feedback from the d.c. side of the convertor to the control system. As explained in Section 5.3.3 if the direct current is distorted, the distortion is fed back and the control system will produce non-equally-spaced firings, giving rise to uncharacteristic harmonics and even direct current in the convertor transformers. The harmonic currents produced by transformer saturation need to be added to the convertor currents and the total current is then injected into the a.c. system/filter parallel combination.

If, in the iterative process, the harmonics at each iteration do not converge and the computed firing instants of the valves show increasing unbalance, this will indicate a practical instability. For instance, if a phase-locked oscillator were used, the instability indicates that the phase of firing would be continuously hitting firing-angle limits, i.e. without reaching a stable operating point. In practice this type of condition would cause higher distortion levels and the system would be disconnected. When these phenomena are investigated in physical models (e.g. the case described in the last section) it is important to represent the system losses and transformer saturation accurately.

5.4 D.C. power modulation

Considerable degree of freedom exists in respect to the levels of power to be interchanged between h.v.d.c. connected systems. This property, although still far from being fully exploited, is used in some existing schemes to provide various degrees of d.c. power modulation which are described in this section.

5.4.1 Frequency control
The frequency of a network interconnected to a larger one by a h.v.d.c. transmission link can be controlled by means of a frequency feedback loop acting on the d.c. link controls, such that the small network draws the required power change from the larger one.

A typical case of frequency control is the Gotland link, as originally designed, which included a synchronous compensator. This network once started could be operated with no other power feed than the d.c. link. Similarly, when the power rating of the d.c. line is comparable with or greater than the rating of the running generators in the a.c. system to which the line is connected, the line terminal can share in the frequency regulation or even perform it unaided.

5.4.2 Power/frequency control

Often a combination of control modes is used, such that the control signal applied to the current controller is normally power, and it does so as long as the frequency remains within the predetermined limits. Outside these limits, frequency control takes over to assist the system in difficulty. However, when the maximum rated transmission power is reached, the frequency controller becomes inoperative. This combination is illustrated in Fig. 5.9.

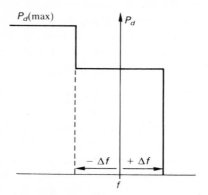

Fig. 5.9 *Power/frequency control characteristic*

It should be remembered that the d.c. link is insensitive to frequency variations, unless some sensitivity is deliberately added to the control system. Without it, a constant power flow can overspeed a receiving system which has lost part of its load and a sending system can eventually collapse if the required d.c. link power is more than the connected generation can produce. In general, therefore, the incorporation of an element of frequency control should be recommended.

5.4.3 Dynamic stabilisation of a.c. systems

A power system is stable if after a disturbance it returns to a condition of equilibrium. This is manifested, not by the constancy of absolute rotational speed of the various machines involved, but rather by these machines swinging together until a new common speed is reached. The power exchanged between them is determined by their relative angular position and therefore, when the equilibrium is disturbed, their rotor positions must give rise to corrective power flow leading to the new state of equilibrium.

If the angle between the machines increases steadily the system is transiently unstable. If the machines fall out of step after a period of increasing oscillations around the equilibrium point the system is dynamically unstable. Dynamic instability is rare in tightly connected systems, which are usually well damped for their characteristic frequencies of the electromechanical swing (between 1 and 2 Hz).

However, when large systems are connected by long relatively weak interties, low frequency swing modes result. The response of the power system controls to the synchronising swings associated with these low-frequency modes can produce sufficient negative damping to cancel the natural positive damping of the system. When this happens, oscillations of increasing amplitude occur.

An example of dynamic instability[10] is the northern and southern parts of the Western US power system, which are connected by the parallel Pacific AC and DC Interties with ratings of 2500 and 1400 MW, respectively. The AC Intertie has a long history of negatively damped 1/3 Hz oscillations resulting from interactions between generators with automatic voltage regulators and system loads. As a result of these oscillations, and because the oscillatory tendency imposed a constraint on the amount of surplus northwest hydro power which could be transmitted to the southwest, a control system to modulate the Pacific DC Intertie was developed.

Damping in the Pacific Intertie is produced by small signal modulation of the d.c. power in proportion to the frequency difference across the AC Intertie. This is accomplished using analogue processing of the AC Intertie power measured at the northern end, to obtain a filtered signal proportional to the derivative of a.c. power at frequencies near 1/3 Hz. This signal is applied, through a $\pm 3\%$ (± 40 MW) limiter, to the current regulator at the northern terminal of the DC Intertie, thus the current setting changes are well within the current margin.

Figure 5.10 shows the result of field tests with and without modulation, wherein series capacitor compensation was first switched in and then bypassed. It is also possible to use a.c. current, rather than power, as the modulation (error) signal. Current is more linear with respect to large swing angles, hence will be more effective when a.c. system oscillations approach stability limits.

Successful operation of d.c. modulation was a key factor permitting uprating of the AC Pacific Intertie from 2100 MW to 2500 MW.

Detecting the effect of the modulating frequency (i.e. the 1/3 Hz harmonic power in the case of the Pacific Intertie) is less effective than detecting absolute phase change directly (a method used in the Nelson River scheme); power measurement provides a signal which levels off near 90° (Section 9.2.1) of the phase difference between the e.m.f.s (i.e. right where most response is urgently needed to try and prevent pole-slipping). In fact in most machine swing problems the net peak survival angle is about 130° and yet, above 90° the power measurement method actually gives a reduced output.

5.4.4 Large signal modulation

While the small signal modulation described in the last section is suitable to maintain the state of equilibrium, it is inadequate for the damping of large disturbances.

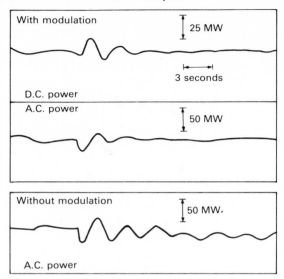

Fig. 5.10 *System response to a.c. intertie series capacitor bypass*

Large signal modulation is thus needed to regain the equilibrium state following large disturbances. A large signal modulation scheme[10] has been added to the Square–Butte h.v.d.c. system in the form of a frequency sensitive power control (FSPC).

A block diagram of the scheme is shown in Fig. 5.11. The frequency deviation (Δf) at the rectifier station is first filtered to eliminate torsional interaction at 11·5 Hz of the 400 MW dedicated generating plant (this effect is discussed further in Section 5.4.6). The output of the stabiliser (ΔI) is limited to 20 per cent of the rated d.c. link current (1000 A); the scheme has been designed to cope with a 20% transient overload.

In order to maintain the current margin between the two ends, the output of the FSPC is also communicated to the inverting station by means of a microwave link using parallel tone channels; each tone channel communicates a 50 A (0.05 p.u.) step change in current order.

5.4.5 Controlled damping of d.c.-interconnected systems
With an a.c. tie line, if one of the interconnected systems is in difficulty following a disturbance, the line is normally tripped to prevent the disturbance affecting the other system, and thus the system in difficulty loses an essential infeed.

An h.v.d.c. link, on the other hand, even with the basic controls, shields one system from disturbances on the other. Although the specified power flow can continue, the option is available to vary the power setting to help the system in difficulty to the extent which the healthy system can allow, without putting itself in difficulty, and subject to the rating on the link.

Interaction between a.c. and d.c. systems 119

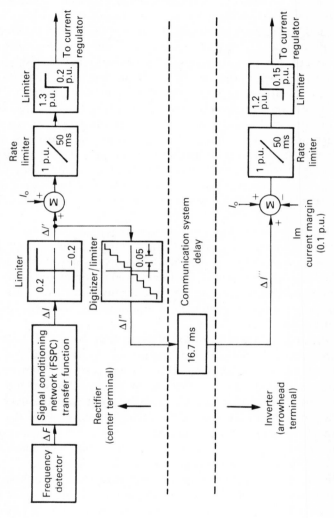

Fig. 5.11 *Square Butte FSPC block diagram*

Although the policy of providing controls which enable the h.v.d.c. link to assist actively in the damping of disturbances should be encouraged, it must be considered that the d.c. link contains negligible energy storage and therefore any action to damp a disturbance at one end must naturally produce some disturbance at the other. In some cases such assistance is readily acceptable, for example when the local system has no directly connected consumers and it can be designed for greater than normal frequency variations. In effect, this allows the inertia of the system to be used to provide the energy for damping the distant system. Another example is where one system is very small as compared with the other, such as the case of an off-shore system where the total load is insignificant compared with the size of the mainland network.

With appropriate control, a disturbance originating in either system can be shared in a predetermined manner, and the resulting system oscillations can be damped simultaneously. Unlike transient stability, where the d.c. link must have the necessary overload capability to get through the first swing, dynamic stability can be achieved without overloading the d.c. link. Moreover if the d.c. link is already operating close to its full capacity, substantial damping can be achieved exclusively by d.c. power reductions at the appropriate instants.[11]

5.4.6 Damping of subsynchronous resonances[12]

The torsional oscillations modes of turbine-generators can interact with the system oscillation modes of the electric power transmission system. With a pure a.c. transmission system, the interaction is dominated by a synchronising component which gives rise to the system modes of oscillation. A relatively small negative damping component is also present, due to the resistance of the a.c. transmission line.

The torsional modes of vibration of the turbine-generator shaft are normally stable when connected to an a.c. transmission system, because of the relatively large positive damping contributed by the damping windings and mechanical damping resulting from steam flow, friction, etc. With the addition of series capacitor compensation in the a.c. network, however, the negative damping contribution of the a.c. system is dramatically increased when the electrical and mechanical resonant frequencies are close. Under these conditions the torsional modes of vibration can become unstable, a phenomenon which is commonly referred to as subsynchronous resonance (SSR).

When the turbine-generator is connected to an h.v.d.c. system, a slightly different situation exists. In the absence of control, the h.v.d.c. system would appear as a load on the turbine-generator which has a positive damping characteristic. In practice, the presence of the current control loop changes the impact of the d.c. system to one of negative damping. This effect can be better understood by considering the case where the d.c. line has no resistance, and the invertor is connected to a system having infinite capacity. The invertor voltage will then be constant, and the use of constant current control will result in a constant power load on the turbine-generator. A constant power load can be shown to have a pure negative damping characteristic. The presence of a parallel a.c. transmission system will decouple the turbine-generator from the influence of the h.v.d.c. system, and will allow the generator damper windings to add more positive damping.

The negative damping contributions due to current or power control with the h.v.d.c. system only occur within the bandwidth of the regulators. Typically, these control functions have bandwidths in the neighbourhood of 10 to 20 Hz. At frequencies above the bandwidth, the d.c. system approaches the situation of no control, which provides positive damping. Hence, only the torsional modes of vibration having frequencies below about 20 Hz will, in general, have the potential for adverse interaction with the d.c. system. This must be treated only as a typical conclusion, however, since individual systems may have wider bandwidths.

This interaction occurred at the Square Butte project through two different control paths; the auxiliary damping control, discussed in Section 5.4.4, and the current control.

In the case of the auxiliary damping control, the interaction was due to high gain and phase lag in the lower range of the subsynchronous torsional frequency region. Thus an effective method of eliminating interaction through this control path was by notch-filtering the auxiliary damping control at 11·5 Hz.

The torsional instability due to current regulator response was more difficult to solve due to its inherent negative electrical damping which cannot be eliminated. However it was possible to minimise the magnitude of the interaction at the frequencies at which the negative damping occurred.

The potential destabilisation of torsional oscillations due to h.v.d.c. systems is similar to that caused by series compensated a.c. transmission lines. However, the interaction with d.c. systems can be solved relatively simply by providing power modulation control to cancel the negative damping impact of the basic constant power control loop. Series compensation, on the other hand, will require special equipment such as series blocking filters and shunt reactive control devices.[13]

5.4.7 Active and reactive power coordination

The degree of d.c. power modulation which can be achieved is restricted by terminal reactive power constraints. With only current or power modulation, an increase in active power transfer will be accompanied by a larger increase in terminal reactive power requirements (see Fig. 2.13) and this effect is particularly noticeable during severe system disturbances. The reactive power variations can cause current control mode transitions between the rectifier and invertor ends, and hence d.c. current changes equal to the d.c. system margin current.

Coordination between the active and reactive power modulation can be achieved by d.c. system voltage modulation. An increase in d.c. voltage will increase the d.c. power transfer as well as the power factor at both terminals, and hence the reactive consumption as a percentage of active power transmitted.

A general a.c.–d.c. system configuration[14] is shown in Fig. 5.12. Each terminal is provided with an independent control loop, shown as a current controller at the sending end and as a voltage controller at the receiving end. Their settings are fixed by the d.c. Modulation Controller, the inputs of which can be either the bus frequencies or a.c. line power flows.

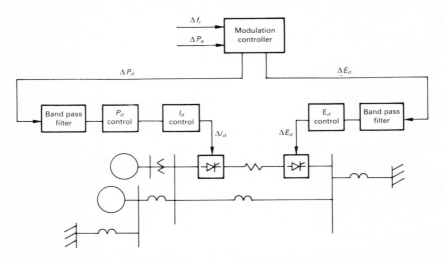

Fig. 5.12 Generic a.c./d.c. system controller structure showing the modulated convertors and a simple interarea a.c. system. Bandpass filters are indicated in each modulation channel

5.4.8 Overall control coordination

The need for an overall control policy is discussed here with reference to the Nelson River scheme. It should be made clear that the specific needs of this scheme and therefore the solutions adopted, may not apply generally. However the overall control philosophy used in the Nelson River project, illustrated in the simplified diagram of Fig. 5.13, helps to understand the degree of dynamic interaction which can be achieved in modern a.c.–d.c. systems.

The following brief description of the main control functions shown in Fig. 5.13 has been extracted from the literature.[15]

(a) The master power controller is designed to accept a power order (P_{01}) from the operator or the automatic generation control. This power order is divided by the measured bipole voltage (V_1) to determine the current order (I_{01}).

Interaction between a.c. and d.c. systems

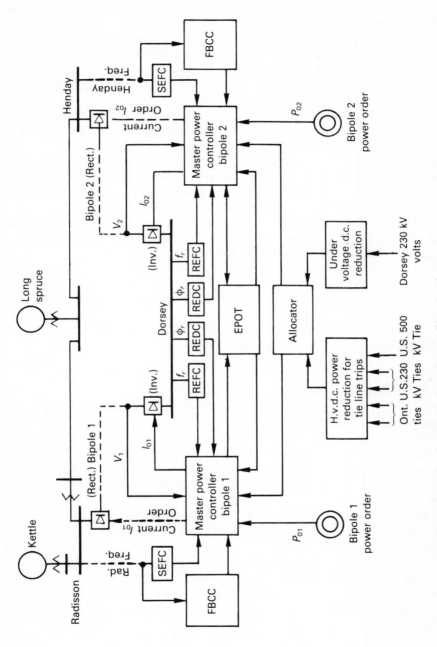

Fig. 5.13 A simplified schematic of controls on the Nelson River h.v.d.c. system

(b) The frequency-based capability control (FBCC) provides protection against h.v.d.c. collector system overloading. D.C. power is reduced if the collector system frequency drops below 59 Hz.

(c) The excess power order transfer (EPOT) controller is designed to fully utilise the combined available h.v.d.c. system capability. The EPOT controller transfers ordered power, which is in excess of the capability of one bipole, to the other operating bipole. This controller is activated only when a major outage occurs on the bipole, such as a valve group or pole block.

(d) The h.v.d.c. power reduction for Tie-Line Trips controller is designed to ensure system stability upon loss of interconnection lines. D.C. power is reduced by an amount equal to tie line loading prior to tripping.

(e) The undervoltage d.c. reduction controller is designed to reduce d.c. power whenever Dorsey voltage starts to collapse. A fixed amount of d.c. reduction releases MVARs both from the h.v.d.c. link and the a.c. system to restore the voltage. This control does not react to faults.

(f) The allocator accepts d.c. reductions from the tie-line reduction and undervoltage reduction controllers. Total d.c. reduction is then summed and allocated to each bipole according to a preset power order.

(g) The sending end frequency control (SEFC) minimises oscillations in the collector system.

(h) The receiving end frequency control (REFC) minimises oscillations in the receiving system.

(i) The receiving end damping control (REDC) operates to prevent changes in the angle of the Dorsey 230 kV bus voltage.

5.4.9 Transient stabilisation of a.c. systems

With the large power ratings of recent h.v.d.c. transmission schemes it is becoming more important to assess the effect of alternative convertor control strategies on the transient stability levels of the interconnected a.c. systems.

While some h.v.d.c. schemes already use special purpose controllers which respond to power-frequency and even absolute phase changes, the possibility of providing d.c. power bursts to reduce first swing stability peaks is yet to be exploited.

The thyristor valves used in h.v.d.c. transmission are rated to withstand considerable overloads without adverse effects to avoid unnecessary protective action. This capability provides the basis for first peak transient stability improvement. The particular strategy, i.e. current increase or decrease, temporary power reversal etc., will vary from scheme to scheme and in each case it can be assessed with the help of a multimachine transient stability programme combined with a small-step transient convertor simulation programme.[16]

5.5 References

1. BOWLES, J. P. (1980): 'Alternative techniques and optimisation of voltage and reactive power control at h.v.d.c. convertor stations', *IEEE Conference on Overvoltages and Compensation on Integrated A.C.–D.C. Systems*, Winnipeg.
2. LE DU, A. (1981): 'Use of static or synchronous compensators in h.v.d.c. systems', *CIGRE Study Committee 14*, Rio de Janeiro.
3. AINSWORTH, J. D. (1967): 'Harmonic instability between controlled static convertors and a.c. networks', *Proc. IEE*, Vol. 114, No. 7, pp. 949–957.
4. YACAMINI, R. (1981): 'Harmonics caused by transformer saturation', *International Conference on Harmonics in Power Systems*, UMIST, Manchester, pp. 102–117.
5. YACAMINI, R. and DE OLIVEIRA, J. C. (1978): 'Harmonics produced by direct current in convertor transformers', *Proc. IEE*, Vol. 125, No. 9, pp. 873–878.
6. OWEN, R. E. (1980): *Study of Distribution System Surge and Harmonic Characteristics*, Section 2.2, EPRI Report, EL-1627.
7. THIO, C. V., McNICHOL, J. R., and McDERMID, W. M. (1980): 'Switching overvoltages on the Nelson River h.v.d.c. system — Studies, experience and field tests', *IEEE Conference on Overvoltages and Compensation on Integrated A.C.–D.C. systems*, Winnipeg, pp. 15–26.
8. AINSWORTH, J. D. (1977): *Core Saturation Instability in the Kingsnorth H.v.d.c. Link*, Paper to CIGRE Study Committee 14.
9. YACAMINI, R. and DE OLIVEIRA, J. C. (1980): 'Instability in h.v.d.c. schemes at low-order integer harmonics', *Proc. IEE*, Vol. 127, Pt. C, No. 3, pp. 179–188.
10. GRUND, C. E., POHL, R. V., CRESAP, R. L., and BAHRMAN, M. P. (1980): 'Increasing power transfer capabilities of a.c./d.c. transmission systems by coordinated dynamic control', *Symposium sponsored by the Division of Electric Energy Systems*, US Department of Energy, Phoenix, Arizona, pp. 371–387.
11. UHLMANN, E. (1975): *Power Transmission by Direct Current*, p. 169, Springer–Verlag, Berlin/Heidelberg.
12. HINGORANI, N., NILSSON, S., BAHRMAN, M., REEVE, J., LARSEN, E. V., and PIWKO, R. J. (1980): 'Subsynchronous frequency stability studies of Energy Systems which include h.v.d.c. transmission', *Symposium sponsored by the Division of Electric Energy Systems*, US Department of Energy, Phoenix, Arizona, pp. 389–398.
13. IEEE SSR Working Group (1979): 'Countermeasures to Subsynchronous resonance problems', *IEEE PES Summer Meeting*, Paper F79 754-4, Vancouver.
14. GRUND, C. E., POHL, R. V., and REEVE, J. (1981): 'Increased performance of h.v.d.c. power modulation by active and reactive power coordination and modern control design', *IEE Conference Publication 205 on Thyristor and Variable Static Equipment for A.C. and D.C. Transmission*, London, pp. 176–181.
15. CHAND, J., RASHWAN, M. M., and TISHINSKI, W. K. (1981): 'Nelson River h.v.d.c. system — operating experience', *IEE Conference Publication 205 on Thyristor and Variable Static Equipment for A.C. and D.C. Transmission*, London, pp. 223–226.
16. TURNER, K. S. (1980): 'Transient stability analysis of integrated a.c. and d.c. power systems, Ph.D. Thesis, University of Canterbury, New Zealand.

Chapter 6
Main design considerations

6.1 Introduction

The designer of a point-to-point h.v.d.c. transmission scheme will normally have a considerable degree of freedom in the selection of design parameters. Before going into detailed specifications it is therefore important to exploit such freedom in order to improve the overall economy.

A typical design sequence should include the following steps:

(*a*) Identify the main operational objectives to be met, i.e. energy considerations, MW loading requirements and maintenance.
(*b*) Identify any technical constraints which may have to be accepted, e.g. the maximum voltage and current ratings of submarine cables, limitations of earth return, etc.
(*c*) Adopt voltage and current ratings.
(*d*) Decide the overall control requirements, e.g. constant power control, short term overload, damping characteristics, constant extinction angle control, constant ideal (no-load) direct voltage, etc.
(*e*) Develop convertor station arrangements.
(*f*) Design the transmission line.
(*g*) Assess the capital equipment cost, the operating costs and the cost of losses.

Steps (*a*) to (*f*) should be critically reviewed to assess the effect of any permissible parameter variation on (*g*).

Although the basic principles of rectification and inversion apply equally to the mercury-arc and thyristor technologies, the design and layout of the convertor plant are greatly influenced by the switching device technology.

The development of the high voltage mercury-arc valve and its influence on the commercial viability of early h.v.d.c. transmission schemes have been discussed in Chapter 1.

Main design considerations

At this point in time the total installed capacity of thyristor schemes has already matched that of the mercury-arc valves, and the latter is expected to be an insignificant proportion of the total by the end of the decade. The design of thyristor convertors is therefore more important from now on and most of the chapter is devoted to it. However, there is still a relatively large number of mercury-arc schemes in existence and some space is also devoted to them in this chapter.

6.2 Thyristor convertors

6.2.1 Thyristor valve architecture

Similarly to the mercury-arc valve, the rating of the thyristor units has been increasing fast. This progress, coupled with the development of thyristor architectures, has led to the present thyristor valve, which is considerably different from the early designs. The progress made in thyristor valves has also had a dramatic effect on the design of h.v.d.c. convertor stations.

Although a concise treatise of the subject cannot do justice to all the individual manufacturers, it would be fair to say that, while there are still small variations in thyristor valve designs, the basic valve and convertor configurations have become more standard in recent years.

The structural design of the modern valve is very simple. It consists of a number of modules in series, each containing several thyristor units. Often, the individual thyristors are rated for the full transmission current with the larger current ratings using no more than two devices in parallel.

A typical thyristor module[1] is shown in Fig. 6.1. It contains six thyristors with their damping and overvoltage protection circuits, voltage dividers and gate control units. The modules are mounted in layers in a case-like structure which constitutes the valve unit.

6.2.2 Twelve-pulse convertor unit

Modern d.c. schemes are designed exclusively for 12-pulse operation. The 'unit concept' of 12-pulse operation is illustrated in Fig. 6.2. In this configuration the convertor station is greatly simplified by the absence of bypass and d.c. isolating switches, which are required for six-pulse operation. The a.c. circuit breaker is common to the two bridges and the fifth and seventh harmonic filters are eliminated.

The four valve units of one arm (or phase) of a 12-pulse convertor configuration are normally stacked together vertically to form a quadruple valve (shown in Fig. 6.3). This arrangement results in the most compact and economical layout of the valves and the valve hall.

Three quadruple valves constitute a 12-pulse convertor. This arrangement can now be designed for any required voltage level. As an example, the three valve group structure of the Nelson River[2] (Bipole 2) scheme is shown in Fig.

design considerations

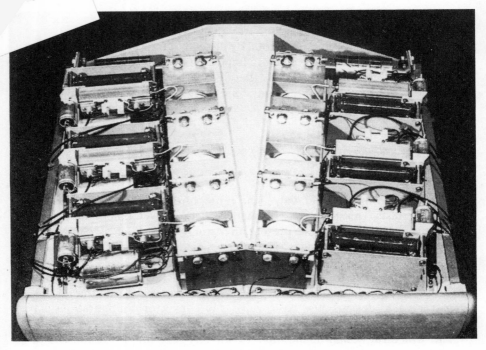

Fig. 6.1 *Thyristor module with six thyristor units (ASEA Journal)*

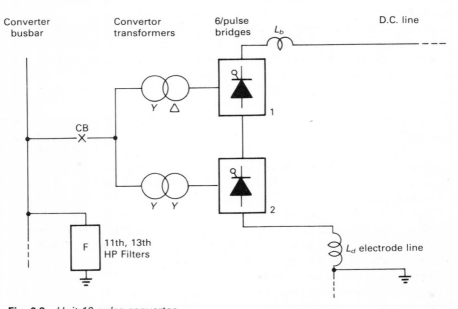

Fig. 6.2 *Unit 12-pulse convertor*

Main design consider

Fig. 6.3 Quadruple valve (ASEA Journal)

6.4. Each group is designed for 250 kV and four such groups in series constitute a ± 500 kV convertor station.

In the case of Fig. 6.4 all the series reactors needed are integrated in the valve structure. Each bridge arm consists of four layers, and each layer (shown in Fig. 6.5b) contains four thyristor and two reactor modules.[3] Therefore the quadruple valve contains 16 layers with a total of 64 thyristor modules and 32 reactor modules. As illustrated in Fig. 6.5a, the valve surge arrestors are directly mounted on the valve structure.

Main design considerations

Fig. 6.4 *Three quadruple valve towers (forming a twelve-pulse group for 250 kV) (Brown Boveri Rev.)*

A cross-section of the 500 kV valve hall and basement is shown in Fig. 6.6. The valve hall contains the quadruple valves with integrated reactors and surge arrestors, surrounded by the necessary clearance for 1300 kV switching surges. The figure also shows the d.c. and a.c. connections. The d.c. busbar connection between the three quadruple valves of a group are made at the bottom, in the middle and at the top. The a.c. connections from one phase are located half way between bottom and middle for the delta (125 kV bridge) and between middle and top for the star (250 kV bridge) connected transformers respectively.

6.2.3 Multibridge convertors

The use of a single 12-pulse convertor per pole, illustrated in Fig. 6.2, for the required full transmission capacity provides the simplest solution, since the a.c. and d.c. system switchyards are greatly reduced. Thyristor convertors with 12-pulse ratings of 560 MW (Inga-Shaba) and 500 MW (Nelson River) are already operating, and there is a demand for ratings of 1000 to 1500 MW per 12-pulse convertor group. There are other considerations, however, limiting the convertor rating. The convertor transformers transport is probably the main one, although the use of single-phase transformer units reduces this problem.

Main design considerations

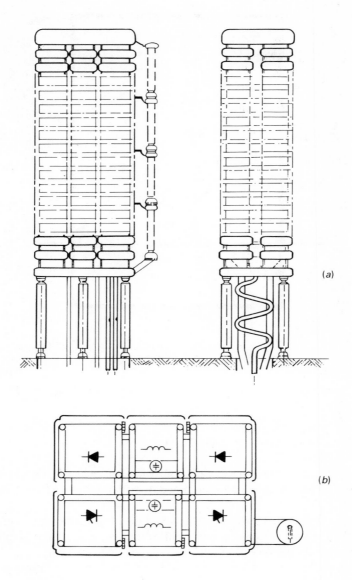

Fig. 6.5 *Quadruple valve*
(a) Mechanical construction, with four convertor arms one upon another
(b) One tier of a quadruple valve

132 Main design considerations

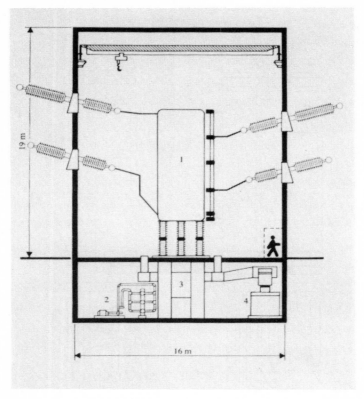

Fig. 6.6 *500 kV valve hall with mechanical auxiliaries (Brown Boveri Rev.)*
    1 = Quadruple valve
    2 = Cooling water system
    3 = Base electronics
    4 = Ventilation of the hall with air filter

In practice, power transmission reliability may often lead to the choice of two 12-pulse convertors per pole, connected either in series or parallel, as illustrated in Fig. 6.7.

These configurations also enable a staged development of h.v.d.c. schemes which often provide more economical solutions, e.g. when a large-scale hydroelectric scheme is continuously developed over the years.

The relative merits of the series and parallel configurations of 12-pulse convertor units depend on the particular h.v.d.c. application. In the parallel configuration of Fig. 6.7(b), a smoothing reactor is required for each 12-pulse convertor and current-balancing control must be used for the two rectifiers or inverters connected in parallel. When the scheme is developed in several stages over the years and the transmission lines only pass through rural areas, the parallel configuration should be used, as it provides less line losses during the

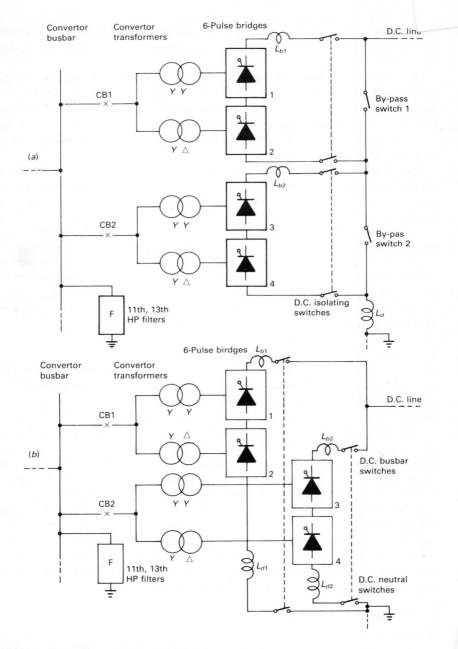

Fig. 6.7 *Double unit 12-pulse convertors*
   (*a*) Double-series unit 12-pulse convertors
   (*b*) Double-parallel unit 12-pulse convertors

period when full line currents are not yet established. This is the case of the Inga–Shaba scheme in Zaire, where the resistive line losses were found to be more significant than the corona line losses at the full line voltages ultimately scheduled.

One of the requirements for the later stages of the Nelson River scheme, and the Pacific Intertie scheme, is that if a transmission line is lost the power will be capable of being transmitted down the other healthy conductors (which will have the required overload capacities), by connecting convertors in parallel. This will result in temporary higher transmission losses, but the availability will be substantially the same as the pre-fault conditions.

Therefore, it is possible to consider high power d.c. transmission systems in which the temporary loss of a convertor unit or the loss of a transmission line will not result in a loss of transmission capacity, at relatively little extra cost to the scheme.

The d.c. transmission must be designed to accept failures within the auxiliary equipment and control systems with only limited loss of transmission capability, a reasonable criterion being that failure of a piece of equipment should not shut down more than one convertor group as far as practicable. The equipment or components, capable of shutting down more than one group following a failure, should be kept to an absolute minimum with redundancy techniques applied. This leads to special design requirements within the control.

6.2.4 Valve cooling system

The thyristors produce considerable heat loss, typically 30 to 40 W/cm^2 (or over 1 MW for a typical quadruple valve), and an efficient cooling system is thus essential. Each thyristor unit is normally provided with a double heat sink and the heat is taken away from the sinks by circulating air or water.

The air is also used for insulation purposes and is thus maintained dry and clean by using a closed-circuit system, including filters and heat exchangers external to the valve hall. Fans are provided in the basement to circulate the cooling air up through a central air duct in the valves, from where it is distributed to all the thyristor heat sinks and voltage divider circuits attached to them.

Water cooling appears to be more advantageous for large power ratings, due to higher cooling efficiency and valve compactness. High purity water combines superb cooling with high electric strength. Water cooling systems, however, require careful design to prevent leakage (which would have disastrous consequences) and corrosion. The water cooling system is normally placed in the basement under the valve hall as illustrated in Fig. 6.6.

6.2.5 Valve control circuitry

Some early thyristor schemes have used magnetic coupling for the firings. In modern thyristor valves all signal communication within, and to the valves, across potential differences is performed using light pulses transmitted by light guides (fibre optics). This applies to both, the firing signals for the individual thyristors in the valve, and the feedback signals from each thyristor level to the valve control equipment.

These feedback signals also make it possible to monitor the state of each individual thyristor. Microcomputers are used in the control room to process the information from the valve. A faulty thyristor is immediately detected and the exact position of the defective thyristor is reported. Since each valve contains a somewhat larger number of thyristors than is actually needed, the convertor can continue to operate, even if some thyristors are defective; these are only replaced during the planned regular maintenance.

The auxiliary power needed for the thyristor firing is obtained from the voltage across the thyristor.

6.2.6 Valve protective functions

Thyristors can be easily damaged if any of their design limits are exceeded; therefore a number of protective functions are normally included in the valve design to override the normal firing control and thus relieve the devices from the overrating condition. The design of the new Cross Channel[4] project valves incorporates the following protective functions:

(a) Current sharing protection (between the two parallel thyristors).
(b) Forward overvoltage firing.
(c) Forward dv/dt protection.
(d) Overtemperature protection.
(e) Forward recovery protection.

The location and basic functions of the Cross-Channel valves are shown in Fig. 6.8.; they are divided into a number of thyristor levels or modules acting independently. As a result, marginal differences in protective settings, or tolerances in valve components, can cause the protective circuits at some levels to operate. This may lead to cascade turn-on, with the last level to fire experiencing a greater duty than that occurring under normal turn-on conditions, and the valve circuit components are rated to withstand such duty.

In the absence of cascade turn-on, the levels that have been protectively fired will conduct the valve grading current. If the disturbance causing such protective level to operate, were now to reverse the valve voltage to a value approaching the protective level of the valve surge arrester (refer to Section 8.8), then those levels which were conducting would be driven by the valve grading current to a prospective negative voltage considerably higher than the reverse voltage rating of the thyristors. However, the thyristors used in this scheme have a high reverse avalanche capability which limits the reverse voltage excursion by conducting the valve grading network current in an avalanche mode, until the valve voltage is more evenly distributed.

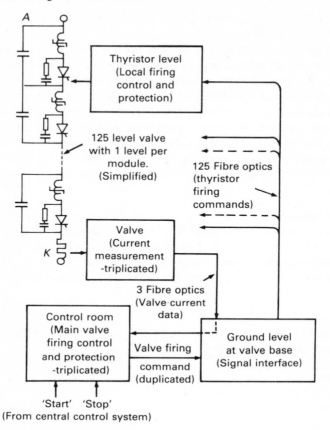

Fig. 6.8 *Location and basic functions of the Cross-Channel valve electronic systems*

6.2.7 Thyristor valve tests

In the past each manufacturer decided on the most appropriate line of action, often determined by the particular customer's requirements.

Test procedures used by several manufacturers and described in a recent CIGRE paper[5] are summarised in this section; an alternative testing procedure, used for the Itaipu project valves is thoroughly described in Reference 6.

An internationally agreed specification (IEC 700) was finally set up in 1981 for the testing of thyristor valves.

Individual components: The individual thyristors are thoroughly tested throughout the manufacturing process. The final assembly of the device with the heat sink is further tested for forward voltage drop, current balance if the design involves paralleling of thyristors on the same heat sink, rate of change of current and gating capability.

Tests specifications also require checking the protective devices (for forward and reverse protection) at this stage.

Each resistor is tested for poor mechanical joints or defective connections; the test stresses the resistor thermally and mechanically simultaneously.

All valve commutating and storage capacitors are required to pass a temperature cycling test of typically 25°C to 80°C.

Reactors and pulse transformers are individually tested for corona extinction at a level at least 1·25 times the maximum repetitive voltage stress expected.

Valve modules: The valve modules contain all the basic components of a valve and can therefore be subjected to the complete testing procedure. This includes voltage and current tests to ascertain that the assembly has been properly manufactured with respect to firing sequence and polarity of the devices, normal voltage stresses and turn-on and turn-off di/dt duties. During these tests the test equipment must be capable of simulating the scaled stray capacitance of the module.

Randomly selected modules are also subjected to special tests such as d.c. and a.c. corona extinction, impulse and switching surges (positive and negative).

Complete valve: All the functions of the valves are tested in their component modules and extensive tests are performed on the complete valve, which includes a current loop test followed by forward recovery.

Probably the most important high voltage tests are the dielectric tests. These are applied to a production valve structure specially mounted to simulate as closely as possible the environmental conditions of the final destination. In these tests the power modules are de-activated by shorting out each power module and removing the series connections. The purpose of these tests is to demonstrate that the valve is insulated for the BIL levels that are coordinated with the station design and the surge arresters.

Also important in the case of air-insulated thyristor valves are the a.c. and d.c. corona tests. The level of corona extinction test must be at least 25 to 30 per cent greater than the maximum repetitive crest value (i.e. the commutation transient with 90° delay angle and at maximum steady state a.c. voltage).

A d.c. high potential test is also carried out on the valve structure, the suggested level being 1·5 times the rated d.c. voltage.

6.2.8 Convertor circuits and components

A single-line diagram of a modern scheme[7] (the CU h.v.d.c. project) is illustrated in Fig. 6.9. In the figure, a cross-stroke on the line indicates a disconnector and a cross a circuit breaker.

The main electrical components of each pole of the convertor station are shown in greater detail in the circuit diagram of Fig. 6.10. All the components enclosed within the thick rectangle are located inside the valve building. The

Main design considerations

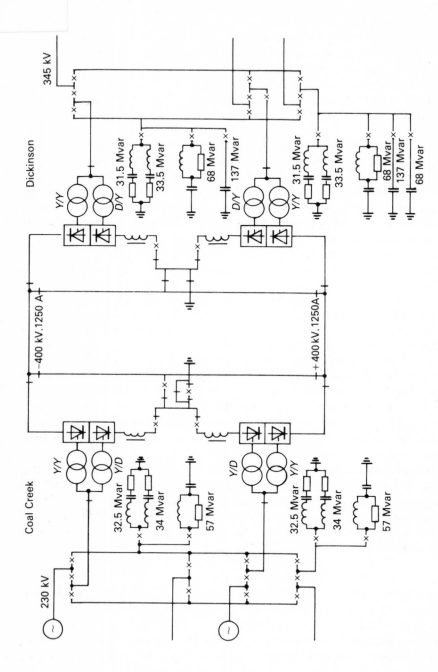

Fig. 6.9 *Single-line diagram of the main circuit of the CU h.v.d.c. scheme*

Main design considerations 139

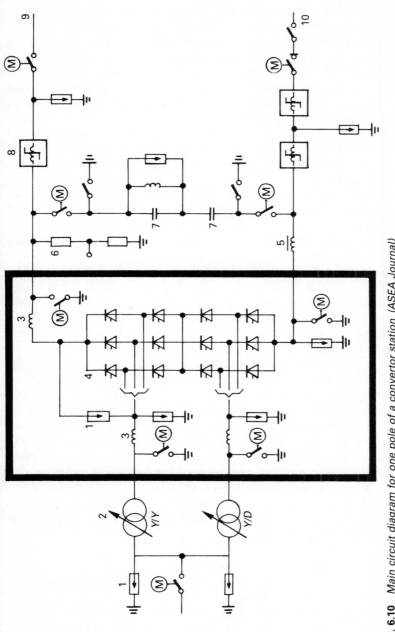

Fig. 6.10 Main circuit diagram for one pole of a convertor station (ASEA Journal)
1. Surge arrester
2. Convertor transformer
3. Air-core reactor
4. Thyristor valve
5. Smoothing reactor
6. Direct-voltage measuring divider
7. D.C. filter
8. Current measuring transductor
9. D.C. line
10. Electrode line

140 Main design considerations

scheme includes two valve groups at each end. Each valve group consists of two series-connected 6-pulse bridges supplied from two convertor transformers. The transformers are connected in star/star and star/delta respectively to provide the necessary 30° phase-shift for 12-pulse operation.

At each end of the link there are two sets of harmonic filters, consisting of tuned branches for the 11th and 13th harmonics and a high-pass branch tuned to the 24th harmonic. The harmonic filters are thermally rated for full bipolar power operation, including continuous and short-time overload factors. A high-pass d.c. filter tuned to the 12th harmonic is also placed on the d.c. side. Extra shunt capacitors are installed at the Dickinson station only, since the generators at Coal Creek can provide the necessary additional reactive power. To limit inrush currents and over-voltages during transformer energisation, the convertor breakers in both stations are provided with pre-insertion resistors.

A d.c. smoothing reactor is located on the low-voltage side and air-core reactors on the line side of the convertors; the latter to limit any steep front surges entering the station from the d.c. side. Additional air-core reactors are installed in each phase on the a.c. side to reduce the rate of rise of current during thyristor turn ON.

The thyristor valves are protected by phase-to-phase surge arresters. The three top valves, connected to the pole bus, are exposed to higher overvoltages in connection with specific but rare incidents, and they are further protected by arresters across each valve. The indoor arrester connected to the low-voltage side of the valve protects the reactor. Pole and electrode arresters supplement the overvoltage protection.

The measuring equipment, i.e. a voltage divider, current measuring transductors and current transformers, provide the necessary input signals for the control and protection circuits.

With reference to switching components (i.e. isolators and circuit breakers), these are of conventional design on the a.c. side of the convertors. Several switches are also used on the d.c. side as indicated in Fig. 6.9. Conventional oil-minimum circuit breakers are used to interrupt small currents for the switching of the neutral bus load and for the change-over from single-pole metallic return to bipolar operation.

Also an h.v.d.c. circuit breaker is used to achieve ground to metallic return transfer (refer to Section 10.2.1); this breaker is designed to interrupt 1500 A and to absorb an energy of 2 MJ.

6.2.9 Thyristor station layout

The area of the modern thyristor station is only a fraction of that needed for earlier mercury-arc convertor stations.

Figure 6.11 shows a typical layout for a 1000 MW bipolar h.v.d.c. station[8] and gives a clear indication of the relative space taken by the various plant components. The major proportion of the space is taken by the external plant

Main design considerations 141

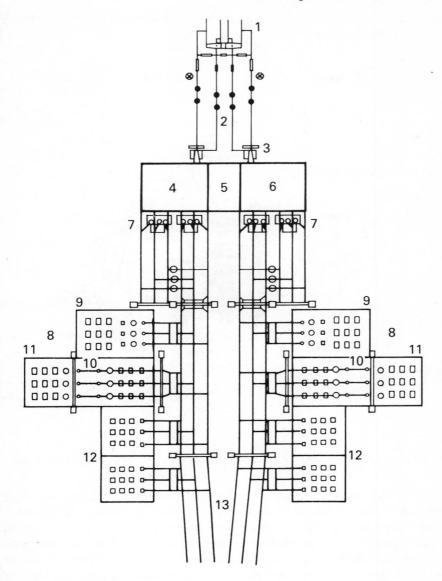

Fig. 6.11 *Station layout for a bipolar h.v.d.c. station (ASEA Journal)*
1. D.C. and electrode lines
2. D.C. switchyard
3. D.C. smoothing reactors
4. Valve hall, pole 1
5. Service building with control room
6. Valve hall, pole 2
7. Convertor transformers
8. A.C. harmonic filters
9. High-pass filter
10. 11th harmonic filter
11. 13th harmonic filter
12. Shunt capacitors
13. A.C. switchyard

components and particularly the capacitors used in the form of harmonic filters and for voltage support.

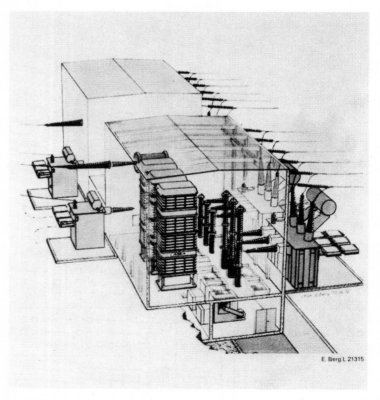

Fig. 6.12 *Sketch of the thyristor valve hall and convertor equipment (ASEA Journal)*

The layout of the valve hall, which apart from the valves contains surge arresters, phase reactors and the line reactor, is better explained with reference to the sketch in Fig. 6.12. The transformers (on the right of the picture) and the smoothing reactor (on the left) are placed close to the hall walls with their bushings passing through the wall. The location of the equipment inside the hall is designed to combine a low probability of internal flashovers with the best possible utilisation of the space available.

The floor area of the valve, service and control rooms is only a small fraction of the total station area. The auxiliary power equipment used for cooling and air conditioning is placed immediately under the valve hall. The building normally contains a steel structure designed to act as a Faraday cage to reduce electromagnetic radiation from the valve hall which might cause radio interference.

6.2.10 Relative costs of convertor components

A recent survey[9] among convertor manufacturers has provided information on costs, losses, overload capabilities and reliability of individual convertor components. The data collection used as a basis a system consisting of two bipoles of conventional equipment. The bipole included four 12-pulse groups, each rated at 800, 1200, 1000 and 2000 MW at a daily mean temperature of 30°C.

The relationship of component costs for a total of 3000, 5000 and 8000 MW ± 600 kV systems is illustrated in Fig. 6.13.

However, it is not practical to use the relative costs in Fig. 6.13 generally, as there are many variables and plant dependencies which are different for each individual application. One of the most important variables in this respect is the ambient operating temperatures, the relationship of rating to temperature, for the more expensive components of the convertor plant, is illustrated in Fig. 6.14.

6.3 Mercury-arc circuit components

6.3.1 Valve group

In mercury-arc convertors the 6-pulse bridge constitutes the valve group, although under normal conditions two phase-shifted groups (12-pulse) operate together on each pole. This arrangement is exemplified by the simplified diagram of Fig. 6.15 which belongs to the New Zealand h.v.d.c. scheme.

Besides the main bridge valves and bridge transformers, the valve group includes a bypass valve, which provides a path for the d.c. current of the series connected bridges during temporary bridge disturbances.[10] For disturbance times exceeding the bypass rating capability a high speed bypass switch is automatically closed across its terminals. Permanent isolation of the valve bridge for maintenance also requires two isolating switches.

The main circuit plant components associated with mercury-arc valves are:

(*a*) Current dividers in series with each of the parallel anodes to achieve current sharing between them.
(*b*) Anode reactors in series with each valve to reduce the rate of change of current.
(*c*) Cathode reactors to damp the high frequency current oscillations due to the commutation process, which would otherwise radiate through the wall bushings.
(*d*) Valve damping circuits in parallel with each valve to control the rate of change of the voltage high frequency oscillations due to the commutation process.
(*e*) High voltage isolating transformers to supply auxiliary power for the excitation system, grid control circuits, vacuum pump, air duct fan and heaters.
(*f*) Surge diverters across the valve group, between phases and from phase to ground.

144 Main design considerations

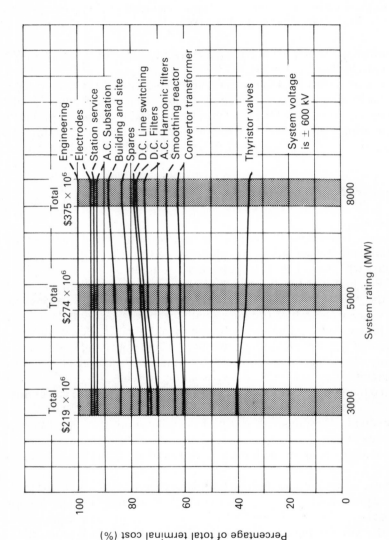

Fig. 6.13 Comparison of h.v.d.c. terminal component costs for basic d.c. scheme with two bi-poles

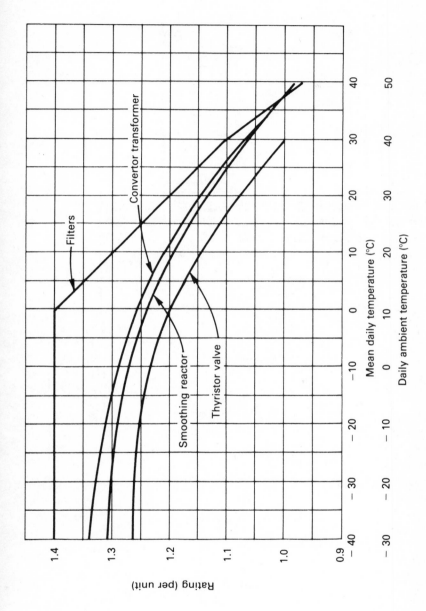

Fig. 6.14 *Relationship of component rating to temperature*

146 Main design considerations

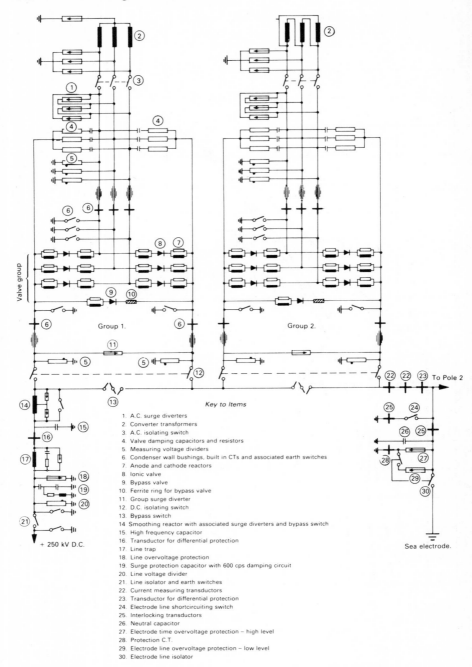

Fig. 6.15 *Elementary diagram of pole 1, Haywards h.v.d.c. invertor station*

(g) More recent schemes also include transient voltage suppression capacitors connected across the valve group to improve the transient performance.

6.3.2 Convertor station

Again with reference to Fig. 6.15, each pole of the New Zealand bipolar scheme includes the following associated equipment in the d.c. switchyard:

(a) A smoothing reactor.
(b) A line surge capacitor.
(c) Surge diverters across the smoothing reactor and between line and ground.
(d) Voltage divider and d.c. current transformers for monitoring purposes.
(e) Some schemes also include 6th and 12th shunt harmonic filters.

The main plant components of the a.c. system illustrated in Fig. 2.7(a) are:

(a) Six single-phase, three winding bridge transformers per pole.
(b) Synchronous condensers connected to the delta-tertiary windings of the convertor transformers. These are needed at the Hayward terminal to supply the extra reactive power required by the convertor and network, and also to increase the sub-transient Short Circuit Ratio of the a.c. system.
(c) A filter bank per pole, which in this terminal includes tuned shunt branches for the 5th, 7th, 11th and 13th current harmonics and a high pass filter for the 17th and higher orders.

6.3.3 Mercury-arc convertor layout

A typical valve house[11] comprises a long, reinforced core building with a steel framed protected metal clad annex on each side.

The central 3-storey core comprises a valve transport corridor with ventilating and air conditioning rooms above and air pressure rooms below in the basement. Along the outdoor station side there are several valve halls with reinforced concrete dividing walls in between and each valve group functions as a self-contained unit with its own ventilating system. Filtered air is drawn into the valve hall via louvres opposite the base of each valve and the air exhausts via roof vents.

The annex comprises a clean work-shop sanitary room, assembly room, degassing room and control room.

Radio disturbance can be transmitted by direct radiation from the valve acting as a dipole. To suppress this, effective screening of the valve hall enclosure is normally required, particularly when the convertor station is close to a residential area. The steel cladding of the outside walls and roof of the valve halls are bonded to provide an effective shielding. A fine mesh screen is embedded in the concrete floors and walls, in addition to the normal reinforcing. Expanded metal screens, well bonded and with phosphor bronze pressure contacts on all gates, are provided between the valve halls and the transport corridor.

6.4 Convertor transformers

The convertor transformers are normally of conventional design. The standard 12-pulse convertor configuration can be obtained with either of the following arrangements.

> 6 single-phase two winding,
> 3 single-phase three winding,
> 2 three-phase two winding.

Star or delta connections are exclusively used for the above configurations. In conventional transformers the insulation gap between winding and yoke is relatively small, as the winding part with potentials close to ground are situated very close to the yoke. In the case of a convertor transformer this is not possible, since the potentials of its connections are determined by the combination of conducting valves at any particular instant, and the entire winding must be fully insulated. As a result of the insulation the radial leakage flux at the ends of the windings increases.

As the leakage flux of a convertor transformer contains very large harmonic content, it produces greater eddy-current loss and hot-spots in the transformer tank.

Apart from the normal a.c. insulation requirement, the convertor transformer is subjected to a direct voltage depending on its position with respect to the ground.

Noise due to magnetostiction is mainly caused by twice the network frequency in conventional transformers. The harmonic content of convertor transformers produces noise to which the human ear is more sensitive, and special measures are often taken like suspending the core, special tank design or providing sound absorbing walls.

On-load tap changing is normally used to reduce the steady state demand of reactive power. The tap change range varies considerably from scheme to scheme, e.g. $\pm 5\%$ in the Sardinia transformers and $^{+17\%}_{-7\%}$ in the CU h.v.d.c. Dickinson terminal.

The tap changer is the most critical mechanical device in use in h.v.d.c. terminals; the number of operations being much greater than in a.c. systems. To enhance the availability of the terminal equipment, vacuum tap changers are used by the GE h.v.d.c. projects in the U.S.

Other important considerations are the shipping weight and dimensional limitations, particularly in modern convertor stations where the valve size is not a limitation any more.

Main design considerations 149

6.5 Smoothing reactors

The presence of a d.c. reactor is an essential part of the h.v.d.c. transmission process. The need for a particular size of reactor, however, does not appear to be a critical factor. While the current waveform, and thus the d.c. harmonic content, improves with increasing inductance, the control response slows down and the resonance frequency reduces making the stabilisation of current control more difficult. The limitation of d.c. line short-circuit currents, to prevent valve damage, is normally the decisive factor and leads to smoothing reactances of between 0·5 and 1 H.

The magnetic circuit of the d.c. reactor consists of an iron yoke enclosing the winding and an air-gap central core. Thus the magnetising characteristic is non-linear and the reduction of inductance at high currents must be taken into consideration when calculating the short-circuit currents.

Thyristor schemes normally use lower values of smoothing reactor because the thyristor valve structures already include intermediate reactors.

Figure 6.2 shows the d.c. smoothing reactor (L_d) on the earth side of the convertor group. This arrangement reduces the overcurrent in conducting valves during earth (flashover) faults within the bridge or at the d.c. terminals; it also allows a reduction of the reactor's insulation requirement. However, an additional small d.c. blocking reactor (L_b) is still required to protect the valves from steep overvoltages, caused by travelling waves originating on the d.c. line or switchyard. The inductance of L_b normally consists of 5 to 10 mH and is of air-core design.

6.6 Overhead lines

The first consideration in d.c. transmission design is the possibility of using the ground as a conductor. Whenever this is possible, a monopolar transmission scheme will provide a very economical solution.

In general, however, the use of ground as a permanent conductor is rarely permitted and a bipolar arrangement is used with equal currents in the two conductors. Normally, if one of the conductors fails, the temporary use of ground return is permitted.

The basic principles determining the dimensioning of overhead lines and hence the towers are very much the same for a.c. and d.c. transmission. Therefore the tower designs for d.c. transmission are very similar to those of a.c. lines. Figure 6.16 illustrates typical towers, used in the New Zealand system. They consist of a ± 250 kV bipolar duplex conductor line; in places close to the sea (case (*d*) and (*e*) in Fig. 6.16) the number of insulator elements is increased to reduce salt pollution problems.

The need for ground wires is less justified in d.c. lines, because (with bipolar lines) only half of the power is affected by a line fault, and the fault current can

Main design considerations

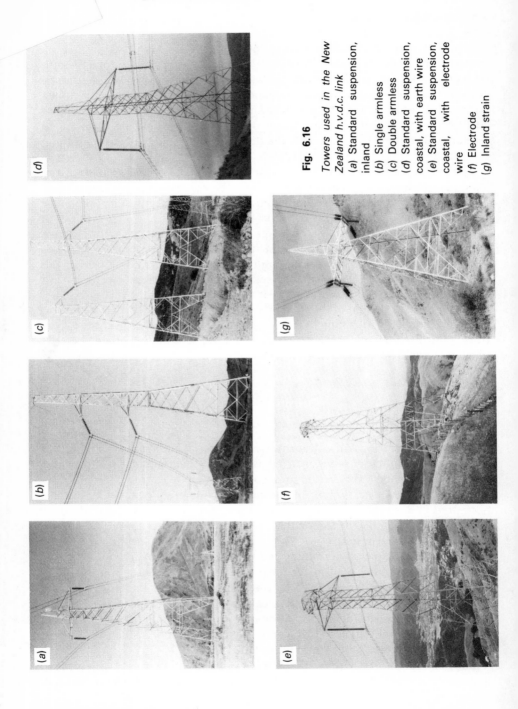

Fig. 6.16

Towers used in the New Zealand h.v.d.c. link
(a) Standard suspension, inland
(b) Single armless
(c) Double armless
(d) Standard suspension, coastal, with earth wire
(e) Standard suspension, coastal, with electrode wire
(f) Electrode
(g) Inland strain

stations are also needed along the route.

be kept under control without the need for mechanical switching. However, ground wires are often used because they relieve some of the harmonic currents flowing through the ground and thus reduce telephone interference.

Considering the independence between d.c. transmission schemes there appears to be no need for voltage standardisation and thus the transmission voltage levels can be chosen to optimise the overall design, i.e. to make the best use of the valve bridges and the overhead line.

With mercury-arc designs the voltages and currents can only be varied in steps, using series connection of valve bridges and parallel connection of valve anodes respectively. With thyristor schemes the voltage can be adapted in a continuous manner; the current rating on the other hand can only be varied in steps, determined by the commercial ratings of the individual devices; to some extent the current rating can also be influenced by the cooling system.

The choice of conductors depends largely on corona and field effect considerations. Results from a recent investigation[12] suggest that those effects do not constitute a special problem for d.c. voltages up to ± 1200 kV. The critical design parameters with respect to corona are radio interference and audible noise, although the corona losses will obviously play a role in the economic choice of conductors. The existence of ion currents under the lines is a characteristic of high voltage d.c. transmission, but the magnitude of touch potentials (from an insulated person touching a ground object) appear to be considerably lower under d.c. lines than those which occur under corresponding a.c. voltage lines.

6.7 Cable transmission

In contrast to overhead line insulators, in which the breakdown can occur through flash-over on the outside, cable breakdown occurs due to puncture through the insulation and this is where d.c. has a major advantage over a.c. Mainly due to the absence of ionic motion in the d.c. cable insulation, the working stress of oil-impregnated paper-insulated cables can be as high as 30 to 40 kV/mm under d.c. compared to 10 kV/mm under a.c. Thus a thicker dielectric is needed with a.c. cables, which impairs cooling of the conductor.

Due to thermal limitations, the power rating of a.c. cables only increases approximately in proportion to the voltage, while the charging current increases with the distance and with the square of the voltage. Consequently, unlike overhead transmissions, a.c. transmission by cable uses relatively low voltages, except for very short distances (mainly on submarine crossings). Also, considerable currents circulate in the sheaths and reinforcing materials, which increase the thermal losses; these losses are often reduced by cross-bonding of cable sheaths, an expensive solution. Forced cooling of the conductors is thus necessary, even at moderate power levels, and cooling stations are also needed along the route.

152 Main design considerations

Moreover, solid insulated d.c. cables can be manufactured for much higher voltages, with considerable reduction in the number of cables needed for a given power rating; this results in less repair problems and wayleave requirements.

Apart from the reduction in losses, resulting from the absence of any appreciable current in the sheaths and reinforcing materials, d.c. cables are subjected to less overcurrent stresses.

Cables to be used under great depths of water need special attention; a typical design is the 250 kV d.c. cable of the Skagerrak scheme laid at a depth of 550 m, illustrated in Fig. 6.17.

A special cable-ship is normally used capable of carrying and laying the complete length of the cable without joints.

Present limitations in d.c. cables ratings may also introduce some restrictions on convertor-station configurations. For instance, the proposed new 2000 MW Cross-Channel h.v.d.c. link between England and France employs two bipolar links, as in Fig. 6.2, with conventional installation of smoothing reactors on the d.c. line side, rather than the multibridge configurations described in Section 6.2.3. The scheme uses eight d.c. cables, rated at 270 kV and 2000 A, buried in four trenches dug into the sea bed.

In shallow waters (e.g. the North Sea) it is recommended to embed the cables into the sea bed (say 0·7 to 1·5 m) to avoid damage by fishing trawls and anchors and thus reduce transmission line forced unavailabilities.

6.8 Earth electrodes

The resistivity of the upper earth layer is typically of the order of 4000 Ωm and, therefore, the electrodes can not be placed directly in contact with earth of such high resistance. Other technical reasons influencing the siting of the electrodes are:

(a) the possibility of the d.c. current ripple interfering with power systems, telephone systems, railways, etc.;
(b) metallic corrosion caused by direct current in equipment in contact with earth such as cable sheaths and pipes.

Consequently the earth electrodes must be placed in an area of sufficient thickness and conductivity and sufficiently distant from urban areas, pipelines, etc.

Kimbark[13] and Uhlmann[14] have given considerable coverage to the experience gathered on the subject of earth resistivity and current distribution in the ground. Such information is used in the design of modern electrodes, such as the one illustrated in Fig. 6.18, which is placed some 22 km away from the Dickinson convertor station[7]; the figure shows one of the 18 vertical rods,

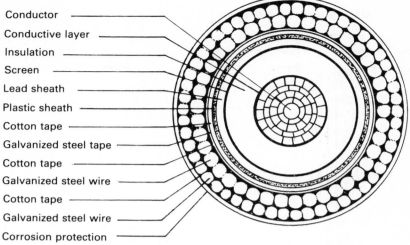
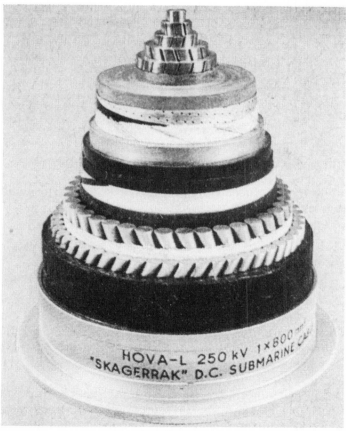

Fig. 6.17 *Cross-section of the double-armoured 250 kV d.c. cable with a cross-sectional area of 800 mm² (ASEA Journal)*

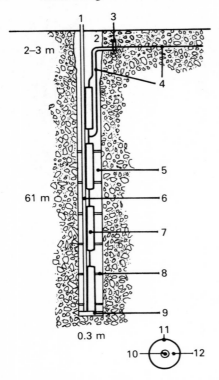

1. Plastic vent pipe
2. Gravel backfill
3. Clad weld
4. Insulated underground cable
5. Coke backfill
6. Perforated PVC vent pipe with a diameter of 38 mm
7. Electrode distribution rod (6 rods per electrode)
8. Centering brace
9. Plugged bottom
10. Electrode rod
11. Casing
12. Coke fill

Fig. 6.18 *Single vertical ground electrode at the Dickinson Station (ASEA Journal)*

which is 300 mm in diameter and about 70 m long. The total electrode resistance is less than $0.1 \, \Omega$.

6.9 Design of back-to-back thyristor convertor systems[15]

The power handling capacity of a thyristor increases with increasing current. For the largest sizes presently used in h.v.d.c. schemes the use of water cooling permits higher current ratings. However, in two-terminal transmission schemes, the rated voltage and current values are mainly determined by the transmission line, which means that it may often not be possible to fully utilize the capacity of modern large thyristors.

In a back-to-back link there are no such external requirements and it is therefore possible to utilize the thyristors optimally, which means a high current and a low voltage. This will minimize the number of thyristors, and thereby also the valve cost. Other voltage-dependent costs will also be reduced, for instance those of the building and the transformers. As an example a 500 MW back-to-back unit will, for a d.c. current of 3000 A, require a direct voltage of only 167 kV.

The circuitry of a back-to-back link does not differ very much from that of two separate terminals. The main difference is that the transmission line is eliminated and, consequently, some d.c. equipment can be avoided or shared by the rectifier and the invertor. Figure 6.19 shows a typical circuit diagram for a back-to-back convertor station.

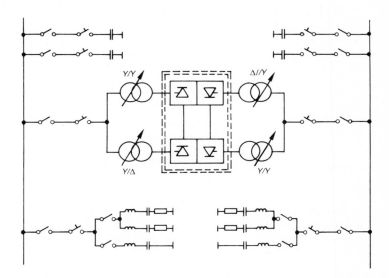

Fig. 6.19 *Basic circuits for a back-to-back convertor station*

For existing air-insulated convertors a practical arrangement has been to stack four valve units on top of each other to constitute one quadruple valve. Three such structures, forming a 12-pulse valve group are placed in a valve hall and connected to transformers placed close to the wall with their bushings inserted into the valve hall through the wall (as shown in Fig. 6.12).

The same arrangement is suitable also for a back-to-back station. With the back-to-back circuitry the two valve halls can be combined into one, with the d.c. loop maintained inside the hall and with transformers on both sides of the building. This is illustrated in Fig. 6.20, which shows a typical layout for a modern back-to-back station. The switchyard is shown for only one of the networks, as the other can be assumed to be equivalent.

As is normal the a.c. switchyard with buses and filters dominates the picture, whereas the convertors themselves occupy only a minor area.

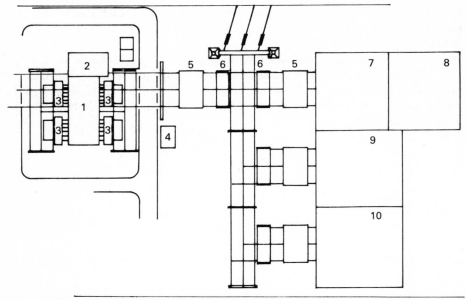

Note: A.C. switchyard with filters is shown for one a.c. system only.

Fig. 6.20 Layout of a modern back-to-back convertor station
1. Valve hall
2. Control and service area
3. Convertor transformer
4. Recooler for valves
5. Circuit breaker
6. Isolator
7. 11th harmonic filter
8. 13th harmonic filter
9. High pass filter
10. Shunt capacitor

6.10 References

1. HAGLOF, L. and HAMMARLUND, B. (1980): 'The Skagerrak transmission — the World's longest h.v.d.c. submarine cable link', *ASEA Journal*, Vol. 53, Nos. 1–2, pp. 3–11.
2. KAUFERLE, J. and KLEIN, H. (1978): 'The compact reliable convertor stations of Nelson River Bipole 2', *Brown Boveri Review*, Vol. 65, No. 9, pp. 578–584.
3. BUELOW, A. F., KAUFERLE, J., and SHEMIE, R. K. (1977): 'Manitoba h.v.d.c. system — Nelson River Bipole 2; General design of convertor terminals', *CIGRE Study Committee 14*, Winnipeg.
4. WOODHOUSE, M. L., BALLAD, J. P., HADDOCK, J. L., and ROWE, B. A. (1981): 'The control and protection of thyristors in the English Terminal Cross-Channel valves, particularly during forward recovery', *IEE Conference Publication 205 on Thyristor and Variable Static Equipment for A.C. and D.C. Transmission*, pp. 158–163.
5. DEMAREST, D. M. and STAIRS, C. M. (1978): 'Solid state valve tests procedures and field experience correlation', *CIGRE Paper 14-12*, Paris.
6. NILSSON, A., EKLUND, L., and HOGBERG, K. E. (1981): 'Design and testing of h.v.d.c. thyristor valve', *IEE Conference Publication 205 on Thyristor and Variable Static Equipment for A.C. and D.C. Transmission*, pp. 154–157.
7. FLISBERG, G. and FUNKE, B. (1981): 'The Cu h.v.d.c. transmission', *ASEA Journal*, Vol. 54, No. 3, pp. 59–67.

8 ERIKSSON, G. and HAGLOF, L. (1975): 'H.v.d.c. station design', *ASEA Journal*, Vol. 48, No. 3, pp. 61–65.
9 HARDY, J. E., TURNER, F. P. P., and ZIMMERMAN, L. A. (1981): 'A.C. or D.C.? One utility's approach', *IEE Conference Publication 205* (as in Reference 4), pp. 241–246.
10 ADAMSON, C. and HINGORANI, N. G. (1960): *High Voltage Direct Current Power Transmission*, Garroway, London.
11 FYFE, R. J., LOUDEN, M. A., NOBLE, J., and YOUNG, D. G. (1966): 'Some features of New Zealand's interisland h.v.d.c. transmission', *IEE Conference Publication 22 on High Voltage D.C. Transmission*, Manchester, pp. 375–379.
12 MARUVADA, P. S. (1980): 'Corona and field effect considerations in the design of UHV d.c. transmission lines', *Symposium sponsored by the Division of Electric Energy Systems*, US Department of Energy, Phoenix, Arizona, pp. 481–505.
13 KIMBARK, E. W. (1971): *Direct Current Transmission*, Wiley Interscience, New York.
14 UHLMANN, E. (1975): *Power Transmission by Direct Current*, Springer-Verlag, Berlin/Heidelberg.
15 CARLSSON, L. and SVENSSON, S. (1980): 'Back-to-back h.v.d.c. links — some aspects on their application and design', *Symposium sponsored by the Division of Electric Energy Systems*, US Department of Energy, Phoenix, Arizona, pp. 199–209.

Chapter 7
Fault development and protection

7.1 Introduction

D.C. convertor stations form an integral part with the a.c. power system, and their basic protection philosophy is thus greatly influenced by a.c. system protection principles.

There are, however, two technical reasons which influence some departure from the conventional protection philosophy, i.e. the limitations of d.c. circuit breakers and the speed of controllability of h.v.d.c. convertors. Furthermore, the series connection of convertor equipment also presents some special problems not normally encountered in a.c. substations.

As with a.c. protective systems, d.c. safety margins should be based on statistical risk evaluations, distinguishing between independent disturbances and the possible cascading of faults. For a given disturbance, the protective system must also be capable of disconnecting only the lowest necessary level and for the minimum time interval.

The characteristics of internal (within the convertor) and external faults are quite different and are considered separately.

7.2 Convertor disturbances

According to the origin of the malfunction, convertor disturbances can be divided in three broad groups, i.e.:

(a) Malfunction of the valves or their associated equipment. The main types are: Misfire, Firethrough and Backfire (or arc-back).
(b) Commutation Failure, the most common disturbance during invertor operation. This fault often follows other internal or external disturbances.
(c) Short-circuits within the convertor station. Although these faults are rare, they must be taken into consideration in convertor design.

7.2.1 Misfire and firethrough

Misfire is the failure to fire a valve during a scheduled conducting period and Firethrough is the failure to block a valve during a scheduled non-conducting period. These faults are caused by various malfunctions in the control and firing equipment.

The effect of these faults is more critical when they occur at the invertor end. With rectifier operation they do not constitute a serious disturbance unless they are sustained, in which case they can introduce voltage and current oscillations on the d.c. side.

By way of example, Fig. 7.1. illustrates the development of a firethrough in valve V_1 at instant B during invertor operation. The valve voltage V_1 is indicated in thick broken-line in Fig. 7.1(b); this valve can firethrough at any time after instant A, although the scheduled firing instant is F. If the cause of the firethrough persists, the fault will recur at instant G, as the thick dotted lines indicate.

It must be pointed out that the idealised waveforms of Fig. 7.1 are only valid in the presence of infinite smoothing inductance and a very large Short Circuit Ratio (refer to Section 5.1). In practice the current will change considerably during the disturbance and, with it, the level of distortion of the voltage waveforms (as explained in Section 7.3).

7.2.2 Commutation failure

This fault is the result of a failure of the incoming valve to take over the direct current before the commutating voltage reverses its polarity, taking into account the need for sufficient extinction time.

A true commutation failure is due to varying conditions in the external a.c. or d.c. circuits combined with inadequate predictive control of the invertor extinction angle. Either a low alternating voltage, a high d.c. current or both, can prevent completion of the commutation process in sufficient time for safe commutation; in such cases the direct current is shifted back from the incoming valve to the previously conducting valve.

Figure 7.2 illustrates the idealised development of a single commutation failure. For simplicity the fault is created by introducing some delay in the firing of the incoming valve, i.e. valve V_3 is fired at instant B, instead of the normal instant A. Since the commutating voltage (phases RY) becomes positive after instant C, the incoming valve V_3 eventually ceases conducting at E (shown by current waveform (e)) and the direct current commutates back to the preceding valve V_1 (waveform (c)).

When valve V_4 is fired at D (waveform (f)), a three-phase short-circuit is briefly established by the conduction of the four valves V_1 to V_4 until instant F, when the commutation from V_2 to V_4 is completed. Between D and I the bridge is bypassed by the conduction of valves V_1 and V_4; during this period there are no alternating currents in the convertor transformer, but the direct current increases in practice as a result of the temporary voltage collapse of the

160 Fault development and protection

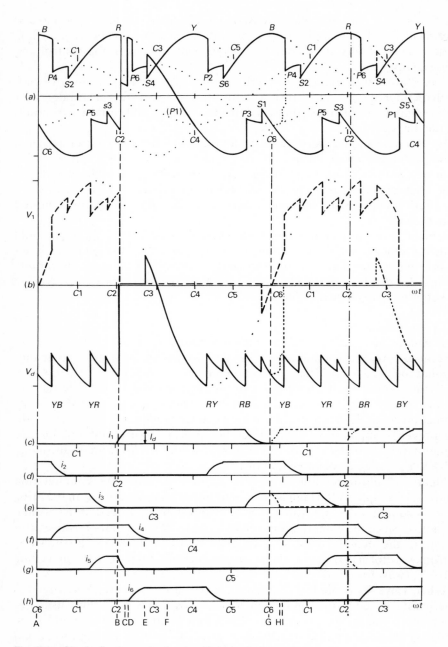

Fig. 7.1 *Single firethrough of valve V_1 in typical invertor*

Fault development and protection 161

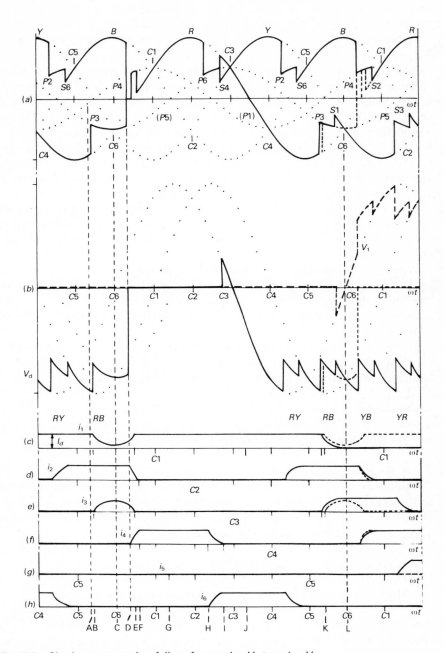

Fig. 7.2 Single commutation failure from valve V_1 to valve V_3

invertor. If a normal commutation takes place from V_4 to V_6 between H and I, the bridge normal voltage is gradually re-established.

However, the single commutation failure, described with reference to Fig. 7.2, is a theoretical occurrence which only happens on infinite busbars. In practice, due to the d.c. current rise, caused by the temporary d.c. voltage collapse at the invertor end, the commutation from valve V_2 to V_4 is also unsuccessful, thus causing a double successive commutation failure. In this case, illustrated in Fig. 7.3, successive valves V_1 and V_2 are the only conducting valves after instant G, and the invertor output voltage V_d reverses for nearly half a cycle (as shown in Fig. 7.3(b)). Such development would, in practice, increase the d.c. current rapidly and, as a result, subsequent commutations may also fail, as indicated by the thick dotted-lines in Fig. 7.3(b). All practical commutation failures are double-successive or worse.

As a result of the d.c. short-circuit at the invertor end, the transformer is either partially or totally bypassed and the d.c. line current exceeds the current in the a.c. lines. This effect has been used to detect the occurrence of commutation failures.

After the occurrence of a commutation failure, the next firing instant is advanced by the constant extinction angle control. If the failure is caused by low alternating voltage following an a.c. disturbance, upon clearance of the disturbance the normal voltage will return and prevent further commutation failures.

In the event of recurring commutation failures the valve group should be blocked. This action, as explained in Section 7.2.5, is often combined with bridge or valve group bypass in the case of a multi-group convertor station.

The probability of commutation failure can be reduced by increasing the minimum extinction angle allowed in normal operation. This, however, increases the VAR compensation required and a compromise is reached where a reasonably low probability of commutation failure is acceptable.

7.2.3 Backfire

Although backfires, or conduction in the reverse direction, can occur (and have occurred) on thyristors, both as external flash-over and as failure of all thyristors in a valve, this fault is only discussed with reference to mercury-arc valves and is caused by the combined effect of:

(*a*) High reverse voltage across the valve;
(*b*) High rate of rise of initial voltage jump;
(*c*) High rate of fall of current at the instant of initial voltage jump.

The reverse voltage is higher during rectification (refer to Fig. 2.3) and therefore the backfire probability is much higher on this mode of operation.

Having lost its unidirectional conduction property, the backfiring valve, together with the forward conducting valve on the same side of the bridge,

Fault development and protection 163

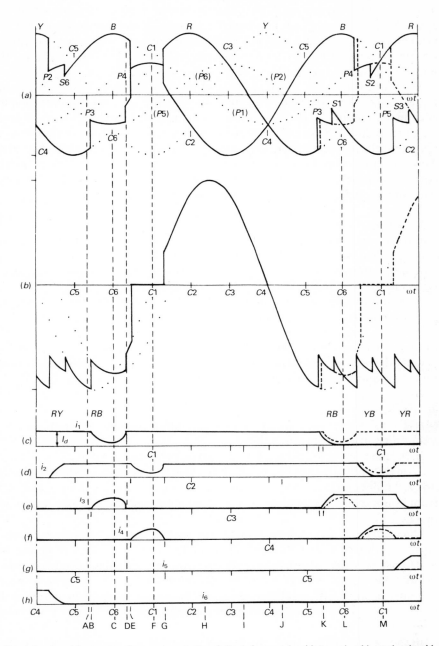

Fig. 7.3 *Double-successive commutation failure from valve V_1 to valve V_3 and valve V_2 to valve V_4*

provides a path for uncontrolled phase-to-phase short-circuit currents in the convertor transformer. Self-recovery is not normally possible with backfires and total blocking is ordered upon detection of a single fault. However, blocking is not always possible and back-up a.c. breaker action is often needed to clear the fault.

The current in the forward conducting valve during this condition reaches typical peak values of 10 p.u.[1] The combination of a high voltage jump following current extinction in the forward conducting valve, and the large current magnitude shortly before extinction (i.e. the ionisation level met by the recovery voltage), often produces what is called a consequential backfire in this valve. This constitutes the most serious condition in mercury-arc rectifiers, as the convertor valves and transformers have to be rated to withstand large overcurrents prior to fault clearance by the a.c. circuit breaker.

7.2.4 Internal short-circuits

Although rare, short-circuits can occur at various locations of the convertor station, as shown in Fig. 7.4. These can be caused by maloperation of earthing switches, deteriorating insulators or surge arrester failures, particularly during transient overvoltages (as discussed in Chapter 8).

A flashover across a non-conducting valve (Fig. 7.4(*a*)) produces a phase-to-phase short-circuit with a very large overcurrent on the conducting valve.

The largest stress is produced during rectification with a small firing delay, and the worst instant is immediately after a commutation, e.g. across valve V_1 in Fig. 7.4; in this case the current in valve V_3 is only limited by the transformer leakage reactance and the system source impedance.

7.2.5 Bypass action

Many of the valve faults are of a temporary nature and can be eliminated by a temporary absence of conduction.

In the mercury-arc schemes this is achieved by the use of a bypass valve across the convertor bridge (see Fig. 6.15). This valve is kept blocked when the bridge unit is conducting in the normal manner. When it is necessary to stop the bridge from conducting, the bypass valve is fired while the main bridge valves are blocked.

Once a bypass valve has fired, it can be blocked only by first interrupting its current so that its grid can regain control. In the case of a rectifier, assuming that its bypass valve is carrying current, when the bridge valves are re-started a positive voltage is established across the bypass valve, the cathode of which becomes positive with respect to its anode; the bridge valves then take over the current from the bypass valve, which stops conducting since it cannot conduct in the reverse direction.

In the case of the invertor, the bypass valve will not stop until its cathode is made positive with respect to its anode; the necessary reversal of polarity may be accomplished by a temporary advance of the angle β to greater than 60°.

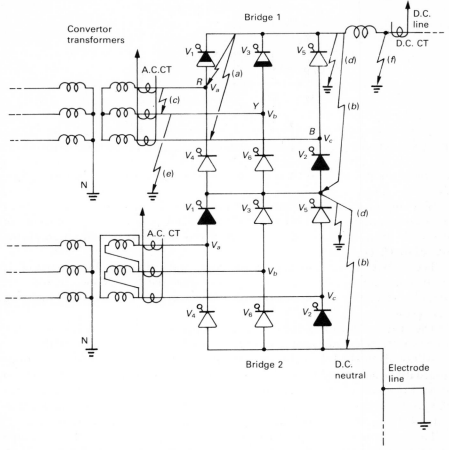

Fig. 7.4 *Possible locations of internal a.c./d.c. short-circuit faults in typical 12-pulse thyristor convertor*
  (a) Faults across a non-conducting valve
  (b) Faults across bridge terminals
  (c) Faults across a.c. phases on the valve side of convertor transformer.
  (d) Ground faults at a d.c. terminal of a bridge
  (e) Ground faults at an a.c. phase on the valve side of convertor transformer
  (f) Ground faults at the station pole or d.c. busbar

A combination of a bypass and two series switches (shown in Fig. 6.15) permits bridge isolation for more permanent outages.

7.2.6 Bypass action in thyristor bridges[2]

The absence of backfires in thyristor valves permits a simpler bypass scheme without the need for a bypass valve. Instead, one of the main bridge arms provides the necessary bypass. A 'healthy' arm pair can always be found to

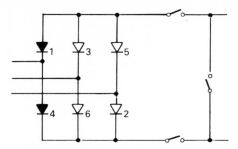

Fig. 7.5 Bridge using a bypass pair (1–4), two series switches and a bypass switch

relieve a temporary disturbance in one of the bridge valves as shown in Fig. 7.5.

Blocking of a convertor through bypass pairs involves the same series of operations, in principle, as blocking through a bypass valve; i.e. blocking of the main firing pulses and the simultaneous injection of continuous firing pulses to a bypass pair.

The selection of blocking sequences applicable to bypass pairs is particularly important to give the faulty valves the best chance to recover, without resorting to the operation of the isolators associated with the convertor.

The criterion for this selection, after a repetitive failure, is that none of the valves of the selected bypass pair should have been involved in the fault.

Although this will be satisfied only by one bypass pair, there are various alternatives according to the selected blocking instant. Ideally bypass action should be carried out immediately after the fault, using the valve which was the last to conduct and its opposite; e.g. if at the instant of the blocking signal, valve 4 is commutating to 6, then the bypass pair to be used should be 3, 6.

The selection of the bypass pair and the blocking sequences are thus simple and few, with the exception of double successive commutation failure; with this fault, two of the valves involved in the fault form a bypass pair, and each one of the remaining two valves belong to either of the two remaining bypass pairs; the simple criterion used to select the bypass pair for a single commutation failure cannot be applied in this case. Generally, however, not all the four valves will be faulty or directly responsible for the occurrence of the fault, and there will be at least one bypass pair through which blocking will be possible.

The selection of the most suitable bypass pair for blocking depends on the cause of the fault. If the fault detectors do not provide sufficient discrimination of the initial cause of the fault and its subsequent development, the selection of the bypass pair and final blocking will be slower.

Resumption of normal operation simply demands the restoration of firing pulses with suppression of the blocking pulses. Invertor deblocking by these means is much simpler than in the case when a conventional bypass valve is used, since momentary rectifier action by advancing the firing angle β is not required.

7.3 Simulation of practical disturbances

The waveforms illustrated in Fig. 7.1–7.3 apply only under idealised conditions in which the convertor fault has no effect on the commutating voltage. In practice, however, such ideal system never exists and even a single commutation failure produces considerable waveform distortion, caused by the temporary harmonic current mismatch between the convertor and the filters. The latter can not accommodate instantaneously the current changes caused by the disturbance, and the difference penetrates into the a.c. system, causing transient voltage harmonic distortion in proportion to the system impedance.

While the convertor is expected to recover from commutation failures caused by normal variations in the a.c. system voltage and direct current, the situation is rather different following large disturbances.

It is thus essential to be able to predict the behaviour of the d.c. link at the design stage and for this purpose the manufacturers make extensive use of scaled-down physical models with detailed representation of the controls. However the losses in physical models can not be scaled down in proportion to the power and, as a result, all the oscillations are subjected to excessive damping; the valve voltage drop is often reduced by means of electronic compensation in the models. Moreover, the so called physical models are also restricted in the number of convertor groups and a.c. system components.

It has been the feeling among some manufacturers in the past that mathematical simulation and computer models could not be trusted to represent the convertor behaviour during disturbances. Such reservations, however, were based on lack of data and the need for large computer requirements. The tendency has been to keep the mathematics simple, while increasing the complexity of the experimental simulators.

More recently the computer requirements, i.e. memory and calculation time, are becoming less important and the mathematical models are getting more sophisticated.[3–6] It appears, therefore, that the main restriction is now the availability of reliable data; a restriction which affects equally the physical simulator.

Some existing a.c./d.c. models, whether physical or mathematical, tend to concentrate on the representation of the d.c. link components, while the dynamic behaviour of the a.c. system plant, and particularly the generators, is oversimplified. Other mathematical models put the emphasis on the a.c. network behaviour and use a simplified equivalent for the d.c. system. Yet, it should be obvious that one without the other will lead to inaccurate prediction of the complete fault development, including recovery.

168 Fault development and protection

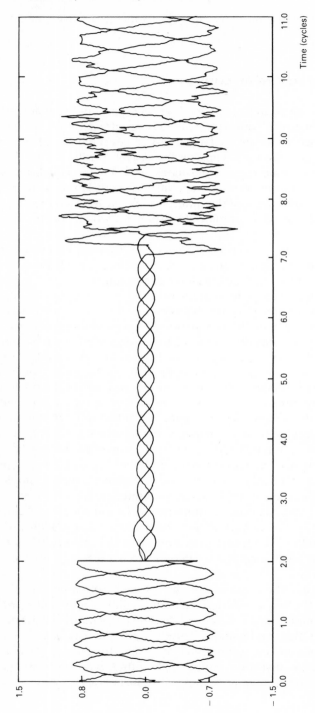

Fig. 7.6 (a) *Invertor a.c. voltages following a 3-phase fault using a time-invariant a.c. circuit model*

Fault development and protection 169

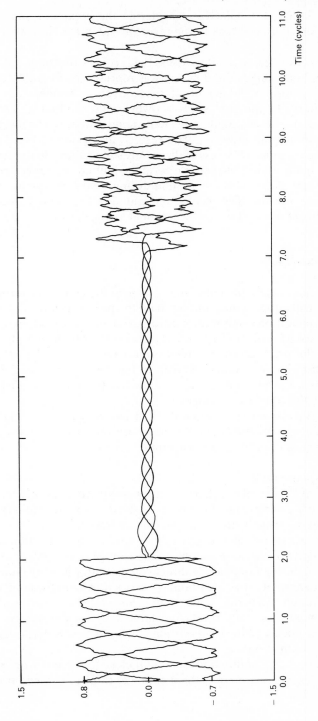

Fig. 7.6 (b) *Invertor a.c. voltages following a 3-phase fault using a time-variant a.c. circuit model*

170   Fault development and protection

Let us illustrate this point by considering the computer simulation of a three-phase short-circuit close to the invertor end in the New Zealand power system.[7]

Two studies were carried out to assess the recovery transient following fault clearance. In both studies the d.c. link was represented in great detail, which included the transient behaviour of the convertor bridges, d.c. lines and harmonic filters. In the first study, however, the faulted a.c. system was represented as a time invariant equivalent circuit, and this was justified on the basis of the short-time involved. The results, illustrated in Fig. 7.6(a) show recovery voltages much in excess of the steady state values. In the second study the dynamic behaviour of the generators in the faulted system was represented in detail, i.e. the equivalent a.c. system during, and subsequent to the fault, was made time variant. The results, illustrated in Fig. 7.6(b), show that the recovery voltages are well below the nominal voltage levels.

7.4 A.C. system faults

Following an a.c. fault, the depressed voltage at the convertor terminals of an h.v.d.c. link will either reduce or eliminate the power transmitted by the link. Under such conditions optimum control strategies must be applied so that normal operation is resumed as soon as possible after fault clearance.

However the d.c. current cannot be reset instantly to the original value because of the time constants involved in the d.c. controls and d.c. circuit. Moreover the a.c. fault will have altered the reactive power requirements and consequently the convertor voltages.

Since the d.c. power transfer consists of the product of voltage (V_d) and current (I_d), any control strategy aiming at fast power recovery needs to take into account both the voltage and current behaviour.

7.4.1 Three-phase faults

The severity of a three-phase short-circuit is greatly reduced as compared with an alternative a.c. interconnection because the d.c. link, due to its fast current controller, feeds virtually no additional current into the fault.

If the fault occurs on the rectifier side no special control action needs to be taken. Provided there is some commutating voltage, the rectifier will continue operating with the highest possible direct voltage and, when the fault is cleared, the rectifier can again recover without special action from the control system.

A short circuit occurring sufficiently close to the invertor end causes commutation failures, thus producing large d.c. current peaks. These are minimised by quickly reducing the firing angle of the valves (i.e. giving more time for valve extinction).

The result of digital simulation[7] of a fault close to the invertor, and cleared after five cycles, is illustrated by the d.c. power transfer behaviour in Fig. 7.7.

Fault development and protection 171

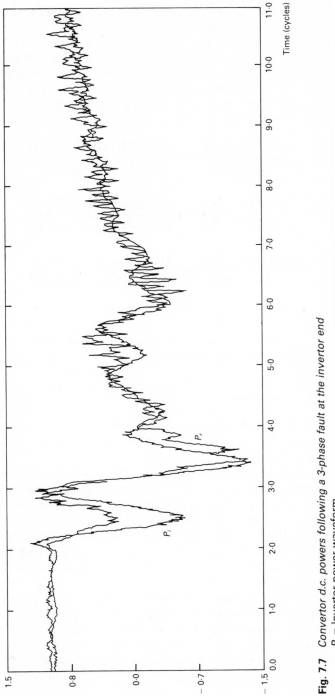

Fig. 7.7 *Convertor d.c. powers following a 3-phase fault at the invertor end*
P_i = Invertor power waveform
P_r = Rectifier power waveform

Fault development and protection

In practice a low voltage limit is applied following an a.c. fault close to the receiving end, as has been explained in Section 4.7.2.

The speed of recovery is a question of optimisation having regard for the a.c. system impedance and voltage and current gradients. Fig. 7.8 illustrates a TNA simulated study[8] of the resumption of power transmission following a three-phase fault at the invertor end. It refers to a back-to-back scheme with a short circuit ratio of 3, and the figure shows the behaviour with optimum gradients of current and voltage.

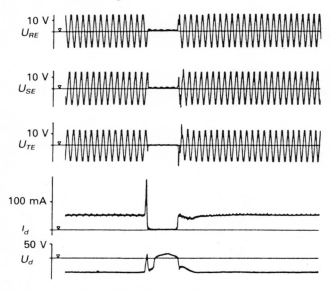

Fig. 7.8 *TNA study of a 3-phase fault at the invertor end*

7.4.2 Unsymmetrical faults

In the case of an unsymmetrical fault there is sufficient commutating voltage for continued operation of the link under reduced power conditions; modern thyristor schemes, however, derive the firing circuitry power from the valve winding a.c. voltage and any protection scheme must take into account that this constraint is met, if operation is to continue. In practice the gate control power supply contains sufficient energy to continue operating for a considerable time during a disturbance, (e.g. three quarters of a second in the case of the New Cross Channel U.K. valves).

The question of re-establishing full power transmission quickly after fault clearance is very similar to the case of a three-phase short-circuit.

Because of the unsymmetry, pronounced double frequency modulation is introduced on the d.c. side[9] which, in the presence of a weak a.c. system, will produce heavy oscillations. In extreme cases, like a line to line ungrounded fault, it may be advisable to interrupt operation while the fault persists.

7.5 D.C. line fault development

The main characteristic of a d.c. line short-circuit is that once started, due to any permanent or temporary fault, it will not be extinguished by itself until the current is brought down to zero and the arc deionised. D.C. faults due to lightning on overhead lines are often self-clearing, because they deionise at the current zeros due to line oscillations; however, this can hardly be guaranteed.

As the fault occurs, the line voltage collapses, the rectifier current tends to rise and the invertor current tends to fall. The invertor will then advance its firing angle, if necessary beyond 90° into rectification (causing a small reversal of voltage on its side of the line) to provide sufficient voltage to maintain the invertor set current.

The rectifier on the other hand will increase its firing delay and maintain its own current setting.

Thus the normal convertor control is not adequate to reduce the fault current to zero; however by suitable control action, the current can be reduced to zero very rapidly as compared with a.c. protection. In practice this is achieved by driving the two terminals temporarily into inversion and thus clear the energy stored in the d.c. circuit faster. This action requires a reversal of the rectifier voltage, following fault detection. The invertor already has the correct polarity, but it must be prevented from going into rectification by setting a limit to its firing angle advance β.

7.5.1 Fault detection[10]

The voltage and current gradients of the travelling waves set up by the fault provide the basis for fast fault detection and discrimination. Their polarity with respect to the line voltage holds sufficient information to identify bipolar and monopolar faults, as well as the poles involved. A monopolar fault can also induce overvoltage on the healthy pole due to mutual coupling.

Fault location and line characteristics affect the overvoltage magnitude at the terminals, but convertor controls have practically no effect on the first wave reflections at the terminals, which in general involve the peak overvoltages.

In the case of a line fault, the rate of fall of voltage at the rectifier terminals is higher than it is for convertor or a.c. system faults, since in the latter case there is much more inductance in the circuit. However, with high resistance ground faults close to the invertor end of a long line, information based exclusively on voltage magnitude and rate of change may not be sufficiently reliable.

For fast d.c. line discrimination it may be better to use the weighted sum of the direct voltage (V_d) and current (I_d) gradients, i.e.

$$\varepsilon = K_1 p V_d + K_2 p I_d$$

which is directly related to the travelling waves initiated by the fault and contains information from which fault type and location can be determined. In a bipolar d.c. line each pole will require this type of detection.

174 Fault development and protection

Although the above considerations have been made in relation to the rectifier end, the invertor end should be equipped with a similar detection scheme (but with different settings) to ensure fast arc extinction.

The sensitivity of the settings of the wave-front detectors has to be assessed by means of actual line tests. For instance in the case of the Nelson River Bipole 2^{11}, and for faults at the remote end of the d.c. line, the steepness values obtained from an early simulation were rather different than those encountered in the actual tests, i.e. 0·3 kV/μs on an h.v.d.c. simulator, 0·7 kV/μs from digital computer simulation and 3·7 kV/μs for the actual plant tests. The digital model has, since, been improved and provides more accurate simulation.

D.C. cable faults are generally permanent and fast detection is not normally used; it is, however, important to provide a very reliable fault detection and location scheme.

7.5.2 Fault clearing and recovery

As indicated earlier, on detection of a d.c. line fault, the rectifier firing angle is delayed into the inverting region (say $\alpha = 120$ to $135°$) to speed up the rectifier current collapse, and is kept at that value until arc extinction and deionisation are likely to be completed. Similarly, to ensure that the invertor end maintains its correct line voltage polarity, it is necessary to limit the invertor firing angle advance (say $\beta < 80°$).

However, the presence of capacitance (particularly with cable transmission) and inductance can create large overswings and polarity reversal of the d.c. line voltage at the invertor end.

On completion of the deionisation period the restart procedure can begin in order to restore normal voltage and prefault power. If re-energisation at full voltage is not acceptable (e.g. due to wet or dirty insulators), then a lower voltage may be used bypassing one or more of the bridges. However, this type of action is not available on the modern 12-pulse schemes.

A starting order is needed to release the emergency control systems of the convertors during the fault. The restart time required will depend on the properties of the d.c. line and the convertor controls. Computer studies carried out in the New Zealand d.c. system[10] indicate that a better performance is achieved by overriding conventional current control and using an exponential function to control the recharging of the d.c. line. Following completion of the deionisation period, the rectifier firing angle is stepped from say $125°$ to $90°$ over one firing instant; subsequent firing action is then controlled by the restart function

$$\alpha_r = \alpha_0 + (90 - \alpha_0).e^{-k.\Delta\theta},$$

where α_0 is the control angle which will give nominal line voltage, $\Delta\theta$ is the elapsed time since the beginning of restart control action (i.e. from when $\alpha = 90°$) and k is a constant controlling the rate of response.

Throughout the recharge period the invertor continues operating under extinction angle control (i.e. $\gamma_i = \gamma_0$).

7.5.3 Overall dynamic response

The result of simulator studies to optimise the fault development of the Nelson River Bipole 2 scheme[11] is illustrated in Fig. 7.9. The fault clearing and recovery action at the rectifier end is controlled by a fixed-rate ramp as can be seen from the trace rectifier α-order. The invertor is clamped at a fixed delay angle during the fault and during restart.

The response of the real h.v.d.c. transmission scheme to a d.c. line fault protection operation (no real fault involved) is illustrated in Fig. 7.10. The d.c. line response is more oscillatory in the actual system, otherwise the comparison of the simulated and actual responses are very similar.

A typical d.c. fault development (relating to the New Zealand system parameters) obtained with a digital model[3] and with the strategy described in Section 7.5.2 is illustrated in Fig. 7.11.

The results illustrated in Figs. 7.9–7.11 clearly indicate that the behaviour of a d.c. system on a line short-circuit compares very favourably with a three-phase short-circuit in an a.c. transmission line. This is due to the presence of the smoothing reactor and the speed of current controllers.

The fault current does not exceed 2 to 3 times rated current, and after about 10 to 20 ms no more than the set current remains. As explained in Sections 7.5.1 and 7.5.2 h.v.d.c. transmission schemes involving overhead line transmission are provided with various d.c. line protection functions, i.e. to detect the line fault, to force the d.c. line current to zero in order to extinguish the earth fault and finally establish a fast recovery.

The d.c. voltage and current are normally restored at controlled rates to minimise recovery transients. The fast control permits very short interruption times in the range of 100 to 200 ms.

7.6 Overcurrent protection

With mercury-arc convertors the highest overcurrents are produced as a result of backfires, particularly when they develop into a consequential backfire in another valve. This situation described in Section 7.2.3, causes overcurrents of typically 10 per unit and often require a.c. circuit breaker tripping to protect the convertor plant, particularly the transformer.

In the case of thyristor-based h.v.d.c. convertors the overcurrent protection settings are decided with reference to the overload capability of the thyristor valves, because of their smaller heat dissipation compared to conventional plant. A typical valve overload capability is shown in Fig. 7.12, which also indicates suggested tripping levels for various disturbances.[12]

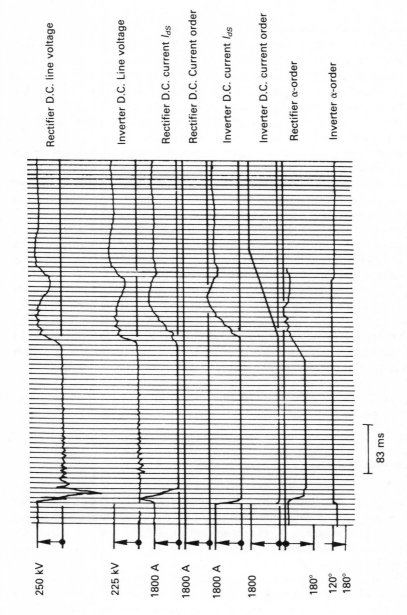

Fig. 7.9 D.C. line fault on an h.v.d.c. simulator after the introduction of an α-ramp in the rectifier and an α-clamp in the invertor (© 1980 IEEE)

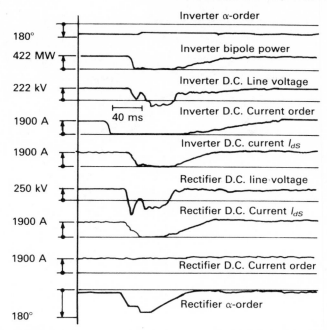

Fig. 7.10 H.v.d.c. system response to d.c. line fault protection operation (redrawn from oscillograms of both substations) (© 1980 IEEE)

The steady state short-circuit current of an uncontrolled rectifier is limited by the increase in overlap angles, which leads to simultaneous commutations on both halves of the bridge and zero d.c. voltage periods. However this situation will not be permitted to develop in practice as the fault current is normally reduced by control action. Of more concern, therefore, is the transient short-circuit behaviour, which, as indicated in Section 7.3, requires a detailed transient simulation of the convertor and a.c. and d.c. systems behaviour. Such studies are necessary to provide adequate overcurrent protection without unnecessary trippings.

Disturbances causing overcurrents in the convertor valves can be divided in three groups; i.e. internal faults, d.c. line faults and invertor-end commutation failures.

Commutation failures may cause long duration overstress if they occur as a consequence of another fault in the receiving end system. Their effect on the rectifier end current is relatively small, since they are separated from the rectifier valves by two smoothing reactors and the d.c. line.

The protection of internal and d.c. line faults is considered next under the subject of valve group and d.c. line protection respectively.

7.6.1 Valve group protection
Internal faults cause severe overcurrents because the impedance between the

178 Fault development and protection

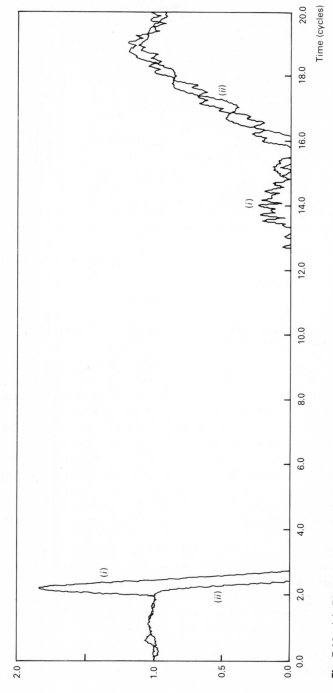

Fig. 7.11 (a) *Direct current waveform during a d.c. line fault*
(i) Rectifier end
(ii) Invertor end

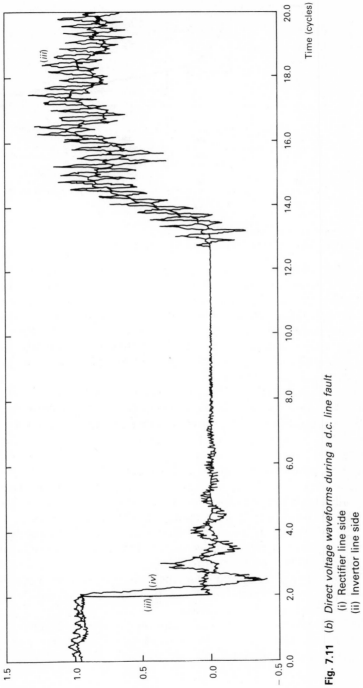

Fig. 7.11 (b) Direct voltage waveforms during a d.c. line fault
(i) Rectifier line side
(ii) Invertor line side

180 Fault development and protection

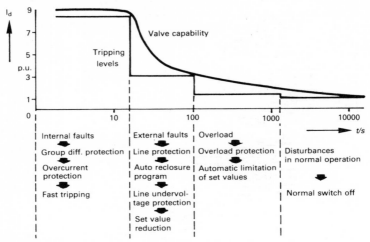

Fig. 7.12 Coordination of overcurrent protection (© 1978 CIGRE)

fault location and the a.c. system source is small. A terminal to terminal short-circuit across one valve, although a rare event, produces the highest overcurrent stress in other valves; typically a current peak of 10 per unit can be expected. Normally, fast suppression of the firing pulses should block the short-circuit current, provided that the valves are capable of withstanding the recovery voltage immediately after the fault. However, if the valve is not able to block, the only way to avoid repetitive overcurrent peaks is by immediate tripping of the a.c. circuit breaker.

A typical[12] overcurrent protection scheme, used for the detection of internal faults in a modern 12-pulse group convertor, is illustrated in Fig. 7.13. The group differential protection compares the rectified a.c. current with the d.c. current. This unit protection scheme provides speed and selectivity.

If the comparison shows that the d.c. is greater than the a.c. current (when the firing angle is in the inverting region) this indicates a commutation failure and the only action required is an automatic increase in the extinction angle γ. However if the condition persists, a temporary block is initiated.

The case of the a.c. being greater than the d.c. current indicates either a backfire (only with mercury-arc convertors) or a short-circuit in the valve group and requires circuit breaker tripping.

A non-unit overcurrent protection scheme is also used as a back-up, with a higher tripping level, to avoid tripping action during faults outside the station which can be cleared by control action.

Finally the detection of ground faults on the d.c. side relies on a pole differential protection scheme, which blocks the convertor valves and trips the a.c. breaker of the affected pole.

7.6.2 D.C. line protection

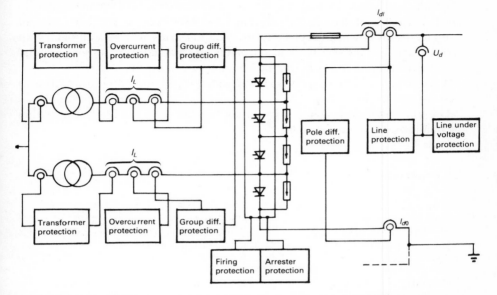

Fig. 7.13 *Overcurrent protection for a convertor station with one 12-pulse group per pole (© 1978 CIGRE)*

D.C. faults are more frequent than internal short circuits and are mainly caused by lightning. The current amplitude is limited by the smoothing reactor and by control action to typically 2 to 3 per unit.

The detection scheme includes rate of change of voltage (which responds after some 3 ms) and a slower back-up line undervoltage relay (which responds after say 50 ms). With parallel lines it is also common to include a rate of change of current comparison scheme to discriminate which parallel line is faulty.

The rate of change is set to discriminate between ground faults and lightning surges not resulting in ground faults; commutation failures at the other end of the line produce less steep voltages at the rectifier end (since there are two smoothing reactors in between) and are also ignored.

A back-up undervoltage unit is also used to detect high resistance d.c. line faults, where the rate of decrease of d.c. line voltage is slow. Although the normal current control can limit the current to a small value, this action is not sufficient to extinguish the fault arc. The fastest way of bringing the line current to zero is to delay the rectifier firing into the inverting region; as a result both stations are temporarily inverting to transfer the energy stored in the d.c. circuit electric and magnetic fields into the the two a.c. systems.

D.C. line protection is not needed in back-to-back interconnections as the two convertors are located in the same building.

7.6.3 Filter protection

The capacitor banks are made up of series connected racks of capacitor units in parallel, each unit having an external fuse. However to prevent frequent prolonged shut downs to replace fuses, the filter arms are also equipped with overcurrent relays. These have an extremely inverse characteristic and respond equally to fundamental or harmonic frequency overcurrents.

A restricted earth fault relay is also provided which operates during an earth fault within any filter arm.

The filter banks are switched by circuit breakers and the filter arms within the banks by disconnect switches.

7.7 References

1 ARRILLAGA, J. and GIESNER, D. B. (1972): 'Recovery of mercury-arc h.v.d.c. interconnectors from backfire faults', *Proc. IEE*, Vol. 119, No. 11, pp. 1611–1615.
2 MORALES, M. (1965): 'Sequential arrangements for the elimination of bypass valves in high voltage direct current convertors', Ph.D. Thesis, Manchester University.
3 ARRILLAGA, J., ARNOLD, C. P., and HARKER, B. J. (1983): *Computer Modelling of Electrical Power Systems*, John Wiley, U.K.
4 DOMMEL, H. W., CHIU, B. C., and MEYER, W. S. (1980): 'Analysing transients in a.c./d.c. systems with the BPA electromagnetic transients program', *IEE Conference Publication 205, on Thyristor and Variable Static Equipment for a.c. and d.c. Transmission*, London, pp. 109–113.
5 HEFFERNAN, M. D., ARRILLAGA, J., TURNER, K. S., and ARNOLD, C. P. (1981): 'Computation of a.c.–d.c. system disturbances — Part I: Interactive coordination of generator and convertor transient models', *Trans. IEEE*, Vol. PAS-100, No. 11, pp. 4341–4348.
6 DISEKO, N. L., WOODHOUSE, M. L., THANAWALA, H. L., ANDERSEN, B. R., DRAWSHAW, A. M., and ROWE, J. E. (1981): 'Application of a digital computer program to transient analysis and design of h.v.d.c. and a.c. thyristor valves', *IEE Conference Publication 205 on Thyristor and Variable Static Equipment for A.C. and D.C. Transmission*, pp. 167–170.
7 HEFFERNAN, M. D. (1980): 'Analysis of a.c.–d.c. system disturbances', Ph.D. Thesis, University of Canterbury, New Zealand.
8 KAUFHOLD, W. and POVH, D. (1981): 'Recovery of the h.v.d.c. transmission after faults in the a.c. system', *IEE Conference Publication 205, on Thyristor and Variable Static Equipment for A.C. and D.C. Transmission*, London, pp. 171–175.
9 GIESNER, D. B. and ARRILLAGA, J. (1972): 'Behaviour of h.v.d.c. links under unbalanced a.c. fault conditions', *Proc. IEE*, Vol. 119, No. 2, pp. 209–215.
10 HEFFERNAN, M. D., ARRILLAGA, J., TURNER, K. S., and ARNOLD, C. P., (1980): 'Recovery from temporary h.v.d.c. line faults', *Trans. IEEE*, Vol. PAS-100, No. 4, pp. 1864–1870.
11 MAZUR, G., CARRYER, R., RANADE, S. T., and WEB, T. (1980): 'Convertor Control and protection of the Nelson River h.v.d.c. Bipole 2 — Commissioning and first year of commercial operation', *Trans. IEEE*, No. 805M6T4-2, Minneapolis.
12 KAUFERLE, J. and POVH, D. (1978): 'Concepts of overvoltage and overcurrent protection of h.v.d.c. convertors', *CIGRE Paper 14-08*, Paris.

Chapter 8
Transient overvoltages and insulation coordination

8.1 Introduction

There are some fundamental differences between the types of overvoltage experienced in a.c. and d.c. transmission schemes, the main differences resulting from

(*a*) The commutation phenomena between the convertor devices, which even during normal operating conditions, results in complex voltage waveforms as shown in Fig. 8.1.
(*b*) The combination of direct and alternating (or transient) voltage.

The bridge units of multigroup convertors are normally connected in series on the d.c. side and in general they are symmetrically placed with respect to earth, as shown in Fig. 8.2. The number of energy-storing elements increases in proportion to the number of bridges on the same side of earth and this results in growing voltage-to-earth stresses on the bridge valves as their separation with respect to earth increases; the stress across each valve is the same, however.

According to their source of origin convertor overvoltages can be divided into:

(*a*) overvoltages excited by disturbances on the d.c. side;
(*b*) overvoltages excited by disturbances on the a.c. side;
(*c*) fast transients produced at the convertor itself;
(*d*) fast transients of external origin (i.e. lightning or switching type surges).

These are first considered separately and their effect is taken into account in the insulation coordination of the convertor station.

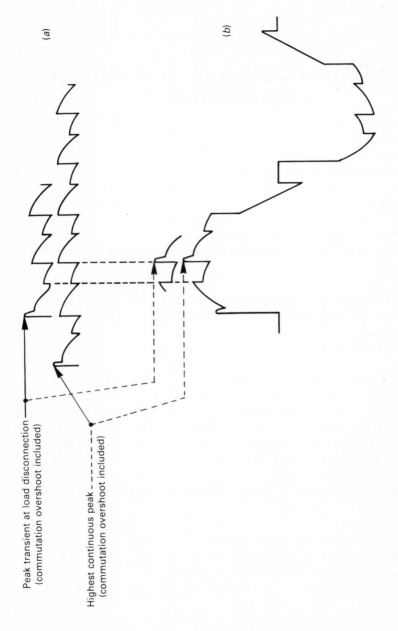

Fig. 8.1 *Voltage waveforms during normal operation*
(a) Voltage across a rectifier bridge
(b) Phase to phase voltage on the valve side of the convertor transformer

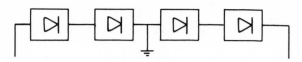

Fig. 8.2 *Multigroup convertor with bridges in series*

8.2 Overvoltages excited by disturbances on the d.c. side

In order of increasing relevance three types of d.c. system disturbance need to be discussed.

(*a*) When a short-circuit occurs in one pole of an h.v.d.c. bipolar transmission system, a transient overvoltage is induced on the healthy pole. However, the level of overvoltage will be limited to under 2 per unit by control action and does not constitute a decisive factor for the insulation coordination.

(*b*) The re-energisation or deblocking of convertor bridges can be a cause of large transient overvoltage. Figure 8.3 shows the overvoltage which can develop on the d.c. line when the rectifier is deblocked with full rectifier voltage, against an open invertor end.[1] However, the start control normally raises the rectifier voltage gradually and the condition shown in Fig. 8.3 can occur only as a result of a fault in the scheme.

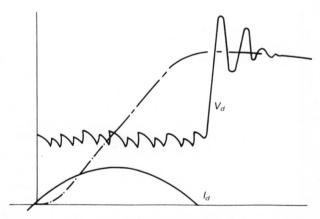

Fig. 8.3 *Deblocking with full rectifier voltage against an open invertor end*

(*c*) The most severe voltage oscillations across a valve will occur in the event, unlikely but possible, of an earth fault on the valve side of the d.c. reactor's bushing; however, this overvoltage does not affect the line insulation. Figure 8.4 shows the maximum rate of rise of calculated overvoltage across the

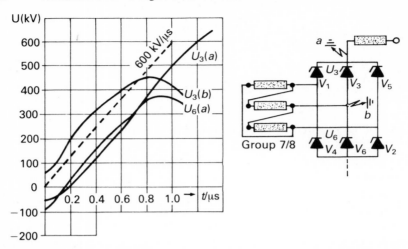

Fig. 8.4 Calculated maximum rate of rise of earth-fault overvoltages across the valves with faults inside the convertor station (© 1974 IEEE)

thyristor valves of the Cabora Bassa scheme[2] with an earth fault at point a or b.

8.3 Harmonic overvoltages excited by a.c. disturbances

The subject of steady state harmonic distortion has been discussed in Chapters 3 and 5. This section considers the problem of transient harmonic overvoltages excited by a.c. disturbances.

The a.c. harmonic filters present a capacitive impedance for frequencies below their lowest tuned frequency. If the a.c. system is inductive, as is the usual case, there is the possibility of a parallel resonance between the filter and the system at one of these low harmonics of the supply frequency (e.g. at the 3rd or 4th harmonic in the presence of 5th harmonic filters), and the lower resonant frequencies will usually result in increased overvoltages.

Following disturbances such as transformer switching, load rejection, a.c. system faults, etc., the resonant circuit can be excited, generating harmonic voltages superimposed on the fundamental frequency, an example of which can be seen in Fig. 8.5.

When transformers are driven into saturation for reasons discussed in Section 5.3.2, they draw harmonic currents and, if the a.c. system exhibits high impedance as a result of resonance at one of these harmonic frequencies, then harmonic voltages of considerable amplitude can be superimposed on the fundamental frequency voltage.

As explained in Section 5.2.1 the most important overvoltages occur during three-phase faults clearance.

Transient overvoltages and insulation coordination

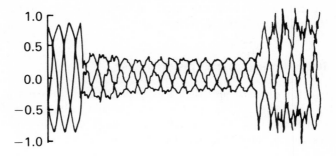

Fig. 8.5 *Harmonic voltage distortion following an a.c. fault*

The level of harmonic overvoltage is very dependent on the effective system resistance, and therefore on the impedance angle of the system, at the harmonic frequency.

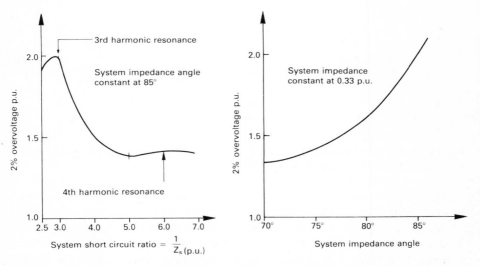

Fig. 8.6 *Statistical overvoltages on transformer energization as a function of system impedance and impedance angle (© 1980 IEEE)*

Figure 8.6. indicates the statistical overvoltages as a function of impedance angle and system impedance, derived from many hundreds of tests carried out on the IREQ h.v.d.c. transmission simulator.[3] The figure illustrates that the lower the impedance angle the smaller the harmonic overvoltages. The presence of local load at a rectifier station therefore tends to reduce the harmonic (transient) overvoltages, in contrast to the increase in regulation (dynamic) overvoltages described in Section 5.2.

Unbalanced a.c. faults cause the injection of second harmonic voltage and current on the d.c. line,[4] which may excite oscillations if the natural frequency of the d.c. line coincides with this frequency.

8.4 Overvoltages due to convertor disturbances

Internal convertor disturbances such as a repetitive misfire or commutation failure in one valve can inject fundamental frequency voltage on the d.c. line. An example of fundamental frequency resonance is shown in Fig. 8.7, which occurred during early operation of the Cabora-Bassa scheme.[5]

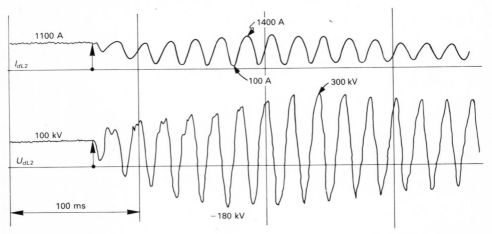

Fig. 8.7 *Line current and voltage recorded at the invertor during missing pulse condition in a rectifier bridge (© 1980 CIGRE)*

Another important source of overvoltage on the d.c. line occurs when the invertor firings are blocked without bypass action. In this case two valves (per bridge) remain conducting and a large a.c. voltage is injected on the d.c. line. Voltage and current oscillations on the d.c. line may then cause current extinction at the invertor followed by a considerable overvoltage, due to the rectifier continuing to provide energy into the d.c. system.

A properly designed convertor current control system is very effective in reducing d.c. overvoltages and oscillations. Even when the line is resonant to the frequency of the injected a.c. voltage, a good controller will limit the overvoltages to less than 50% of the line voltage.

8.5 Fast transients generated on the d.c. system

Under this category we include the following:

(*a*) Fast front surges of external origin and of shapes similar to the lightning

surge. The front and tail times of the lightning surge tend to fall within reasonably closely-defined limits; for most purposes it is taken that a voltage surge associated with lightning may be represented by a unidirectional double-exponential wave, rising to crest from zero in about 1 μs, and falling to half the amplitude in 50 μs. This is the so-called 'standard lightning impulse', or 1·2/50 μs wave.
(b) Slow front surges with shapes similar to the switching surges, e.g. with 250 μs front and 2500 μs tail

The characteristics of lightning and switching surges are well documented and are only considered here in as much as they affect the insulation coordination of the convertor plant.

8.5.1 Lightning surges

The steep wavefront of a lightning surge travelling along the d.c. overhead line is first slowed down by the line surge impedance. The propagation of the incident surge into the convertor station depends on the characteristics of the smoothing reactor and d.c. line surge capacitor.

When calculating the effect of lightning surges the d.c. line surge capacitor is normally replaced by its capacitance and an associated inductance (representing the connections inside the capacitor and to the line); the d.c. reactor coil behaves as an inductance in parallel with an effective interturn capacitance as shown in Fig. 8.8.

By way of example the effect of a 5 kA incident surge reaching the convertor termination is illustrated in Fig. 8.8, using the rise time τ as a variable parameter.[6] These sharp voltage oscillations can be reduced substantially by providing ground wires in the vicinity of the convertor station combined with low tower footing resistance. For a typical increase of surge front time of 1 μs/km, Fig. 8.8 shows that the provision of a 2 km long ground wire can reduce the rise time of the surge across the surge capacitor to 2 μs.

8.5.2 Switching-type surges

Owing to capacitive and inductive coupling between the conductors, earth faults on one pole give rise to surges on the healthy pole similar to those caused by switching (i.e. long front and long tail).

For the analysis of surge transfers to the convertor side the stray inductance of the surge capacitor and the shunt capacitance of the smoothing reactor can be neglected because of their small time constants (relative to the wavefront time).

Transient Network Analyser tests[7] indicate that the peak can reach up to 1·8, depending on line distance and termination.

For a convertor in its normal conducting mode, the distribution of the switching surge is governed by the amount of inductance existing between the points concerned and ground.

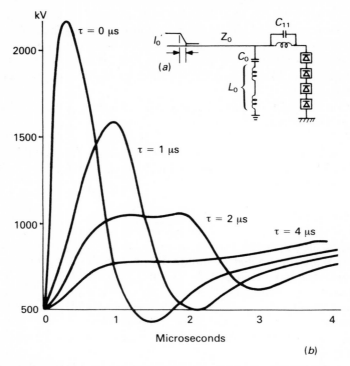

Fig. 8.8 Lightning surge $I_0 = 5$ kA propagating into the convertor station
(a) Main circuit scheme
(b) Surge capacitor voltage
Parameter τ: Rise time of the lightning surge on the d.c. line before the surge is influenced by the surge capacitor
Surge capacitor $C_0 = 0.1$ μF
Surge capacitor connections $L_0 = 100$ μH
Stray capacitance across the smoothing reactor
$C_{11} = 1000$ pF
d.c. line wave impedance Z_0

The voltage across both the upper and lower bridges are mainly determined by the voltage drops across the conducting phases of the transformer belonging to the bridge in question.

Switching surges of long duration and high energy content can cause current levels of sufficient magnitude to produce current extinction[8]; this is illustrated in Fig. 8.9(b). Since the inductance of the smoothing reactor is much larger than that of the convertor transformer leakage, the incoming surge has little effect on the bridge voltages while the valves are conducting. However, when the incoming surge extinguishes the valve current, the station side of the smoothing reactor is connected to earth through the valve grading, damping circuits and stray capacitances only; this causes an overswing of higher value

Transient overvoltages and insulation coordination

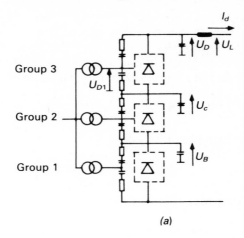

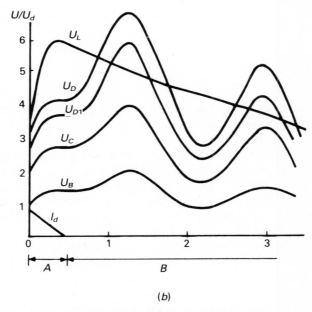

Fig. 8.9 D.C. line switching-surge causing valve current extinction
 (a) Circuit scheme after valve current extinction
 (b) Current and voltage
 A. Inductive voltage distribution between bridges
 B. Capacitive voltage distribution between bridges
 I_d Line current
 U_d Direct voltage/bridge
 U_B, U_c, U_d Voltage to ground from d.c. busbars
 U_{D1} Voltage to ground from a.c. bus group 3
 U_L Voltage to ground on the d.c. line

than the incoming surge. Figure 8.9 shows the overvoltage distribution for three bridges and the top transformer connections.

It is worth pointing out that overvoltages are often the result of certain sequences of events occurring in practice; caution must therefore be used in using purely speculative theoretical studies with possibly incorrect assumptions.

8.6 Surges generated on the a.c. system

Switching surges generated in the a.c. side are inductively transferred to the valve side via the transformer. Since the valve group voltage only consists of phase-to-phase components, the zero-sequence is not included in the convertor voltage.

A single-phase superimposed surge is transferred to the valve side in different ways in the $\lambda - \lambda$ and $\lambda - \Delta$ connected transformers of a two series-connected valve group, the total peak voltage across the group being given by[6]

$$V = n\left[V_m \sqrt{2}\sqrt{2 + \sqrt{3}} + V_0 + 2\frac{V_0}{\sqrt{3}}\right],$$

where n is the transformer turn ratio $(\lambda - \lambda)$, V_m is the maximum permissible steady state operating phase voltage, and V_0 is the peak value of the superimposed switching surge.

The convertor stations and transmission lines in their vicinity are normally shielded to prevent direct lightning strokes. Strokes distant from the convertor station are attenuated by the time they reach the convertor station. Moreover, the steep front overvoltages generated during lightning are subjected to a considerable front prolongation due to the capacitive termination. The tuned a.c. filters and capacitor banks have considerable damping effect on the incoming wave; most of the filter damping is due to the high-pass branch. At low load, however, a number of capacitor banks may be disconnected but the maximum probable overvoltage factor is not likely to exceed 2·2 p.u.

Lightning surges in the a.c. system can be transferred to the d.c. side via the transformer electrostatically and electromagnetically. Generally the electrostatic transfer can be neglected since only waves with a steep wave front are transferred via the stray capacitance.

An example illustrating the capacitive and inductive transformation of a lightning surge is shown in Fig. 8.10. The voltage increase after 60 μs it is seen to depend upon electromagnetic transfer.

However, when the operating voltage in the a.c. system side of the convertor transformer is much higher than the corresponding voltage on the d.c. side, the capacitive transformation may be of importance; for transformer ratios larger

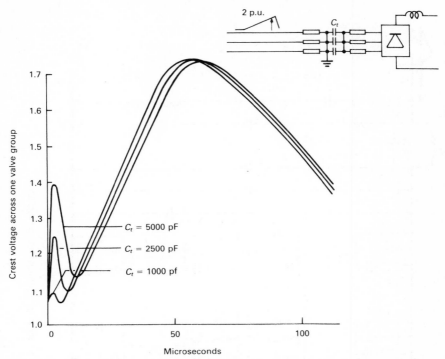

Fig. 8.10 *Capacitive and inductive voltage transformation through the convertor transformer of a lightning surge with the amplitude 2 p.u. superimposed on the operating voltage on one phase*
  Parameter C_t: Capacitive coupling between the valve and line windings of the transformer
  Transformer ratio 1 : 1

than 4 the capacitively transferred component is larger than the electromagnetic.

8.7 Fast transient phenomena associated with the convertor plant

Comprehensive computer programmes are required to assess the internal distribution of transient voltage stresses within the convertor upon the arrival of surges of either internal or external origin.

The physical behaviour and circuit configuration of thyristor and mercury-arc convertors are very different in this respect and are thus considered separately.

8.7.1 Mercury-arc convertors

With reference to the general model of a 12-pulse mercury-arc convertor

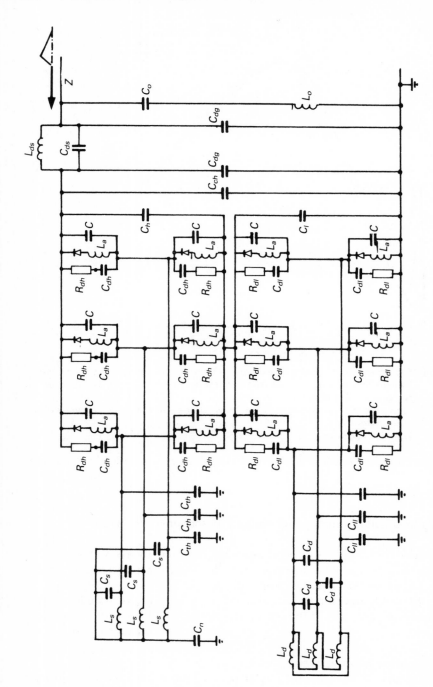

Fig. 8.11 General equivalent circuit for surge-phenomena studies in mercury-arc convertors

suitable for fast transient studies, shown in Fig. 8.11, the following criteria apply to the energy storing elements during operating conditions:

(a) The anode to cathode self-capacitance valve is only effective when the valve is not conducting, the self-capacitance of transformer windings is only effective in the idle phase, and the cathode to neutral capacitance is effective at the start and end of every commutation.

(b) The capacitor, damping circuit and anode reactor associated with the valve exist or disappear according to the three different states of the valve, i.e.

> 'conducting' — all of them are short circuited,
> 'starting conduction' — all of them exist,
> 'stopping conduction' — only the anode reactor disappears.

Moreover, the transient response of the passive circuit depends on the initial values of the instantaneous nodal voltages and injected nodal currents at the instant when the discontinuity takes place.

To represent the different operating conditions encountered in practice, the general circuit must be modified by eliminating or short-circuiting some of the branches.

By way of example, the propagation of a 1000/1/23 incident surge across the 0·5 H smoothing reactor, and the internal overvoltage distribution in a typical double-bridge mercury-arc convertor,[9] are illustrated in Fig. 8.12.

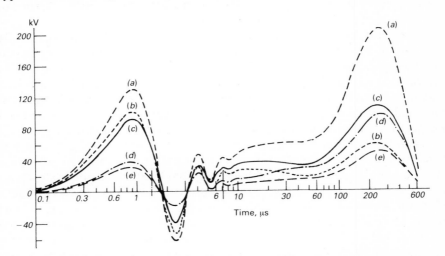

Fig. 8.12 *Convertor voltage distribution for a 1000/1/23 incident surge (nonconducting convertor)*
- (a) On valve side of d.c. reactor
- (b) Across upper valve of upper bridge
- (c) Across upper bridge
- (d) Across lower bridge
- (e) Across upper valve of lower bridge

The equivalent circuit of Fig. 8.11 can also be used to optimise the design of the damping circuits used to limit commutation oscillations. This is carried out by performing a large number of studies, representative of the different types of normal and abnormal commutations,[10] with delay angles producing the highest voltage jump in each case.

For the analysis of arc-quenching phenomena,[11] the faulty valve is simulated by an open switch shunted by the valve stray capacitance in series with the anode reactor represented by a parallel combination of inductance, resistance and capacitance.

8.7.2 Thyristor convertors

The modelling of transient phenomena in thyristor convertors must take into account the thyristor non-linearities such as recovered charge, leakage and displacement current phenomena associated with the various thyristor operating states and the diversity which exists in those properties between the many thyristors involved in the valve structure. These characteristics interact in a complex manner with the other components like the saturating inductors, transformers, damping and grading circuits, stray capacitances and inductances of the valves, busbars, transformer windings, etc.

A comprehensive computer programme[12] is needed as a tool to assess the internal distribution of transient voltage stresses and other related information such as cascade turn-on, overvoltage limitation, protection and coordination, valve recovery at turn-off, voltage unbalance along a series-connected string of thyristors, and transient overvoltage disturbances.

An application of such a model is now described which relates to the cascade turn-on in thyristor valves. If, in a valve equipped with independent overvoltage firing at each voltage level, a component failure causes one level in the valve to rely on this protection for triggering, abnormal voltage excursion and inrush current in excess of the normal will be imposed on the afflicted level. Since this operating regime can persist repetitively until the next scheduled maintenance, it is of crucial importance to valve component ratings.

When the levels in a valve fire non-coherently, the voltage across the last level to turn on will rise at a relatively fast rate. The overvoltage protection gates the thyristor when its switching threshold level is exceeded. However, because of the turn-on delay of the thyristor, a finite time passes before a thyristor impedance falls sufficiently to establish a safe conduction path. The digital programme can then be used to ensure that during this interval, the thyristor and other components are not overstressed due to excessive current, voltage or rate of rise of voltage, and that all components are adequately rated.

A simplified equivalent circuit used to study the cascade firing is shown in Fig. 8.13 for a valve employing n levels with saturating inductance distributed equally between the levels. The circuit external to the valve represents the bridge in respect of valve inrush current over the period of interest. The valve is

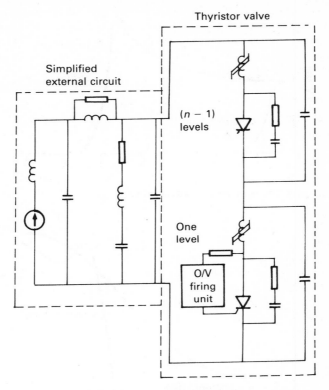

Fig. 8.13 *Simplified equivalent circuit for cascade turn-on investigations*

split-up into two parts, one part representing the late firing level and the other simulating the rest of the $(n-1)$ levels; the impedances of the $(n-1)$ levels are taken with minimum tolerance and those of the late firing level are at maximum tolerance (this maximises the voltage on the last level to turn on).

The equivalent circuit of Fig. 8.13 is initially charged to the appropriate voltage level. At the beginning of the calculation the $(n-1)$ levels turn on simultaneously, with their thyristors represented by time-dependent resistors. During this time, the thyristor in the last level to turn on is simulated by a voltage-dependent capacitor. When the voltage across the thyristor reaches the protection threshold, the thyristor is gated to turn on after a specified delay. The impedance of the thyristor begins to fall in a manner determined by its time-dependent switching characteristics, the sequence being initiated when the gate current attains a preset level.

Typical waveforms of thyristor voltage and inrush current are shown on Fig. 8.14 for the last level to fire. Corresponding waveforms for normal coherent turn-on are also shown for comparison.

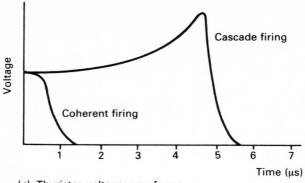

(a) Thyristor voltage waveforms.

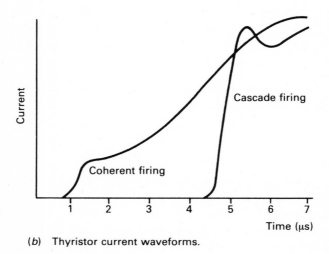

(b) Thyristor current waveforms.

Fig. 8.14 *Typical thyristor voltage and current waveforms for normal and cascade firing*

8.8 Insulation coordination

The generic purpose of insulation coordination is the selection of the most economical combination of plant insulation and overvoltage protection to ensure satisfactory performance of all the insulation around the convertor plant. More specifically, the subject is concerned with the need to protect the convertor plant components against large transient overvoltages for which they cannot be economically designed or compensated.

The limitation of overvoltages, essential to economic power plant design, is carried out in two ways.

Transient overvoltages and insulation coordination

(*a*) By suitable system design.
(*b*) By suitable coordination between insulation and surge arrester protection.

8.8.1 System design
On the d.c. side, overvoltages can be reduced by various means, i.e.:

(*a*) shielding the convertor station and transmission lines;
(*b*) suitable design of the convertor control equipment;
(*c*) use of damping circuits;
(*d*) selecting system parameters to try and avoid resonance under fault conditions

On the a.c. side, voltage support equipment is essential to provide voltage control during transient and dynamic system disturbances. The type and amount of compensating equipment requires detailed studies to determine the relationship between d.c. system recovery, (i.e. changes in convertor control mode, rate of ramping of the d.c. current on restart and stabilising signals) and voltage support equipment rating and response times.

The different characteristics of each type of voltage support equipment have marked effect on the performance of the a.c.–d.c. system which must be studied and evaluated. For example, an a.c. system fault may result in sufficient voltage reduction and distortion to cause commutation failures at the invertor. With synchronous condensers, as they contribute to the short circuit level, such problem will be less likely than with static VAR systems (SVS).

The occurrence of commutation failures during conditions of low or distorted a.c. busbar voltage, either during a fault or on recovery, requires realistic representation of the convertor valves with their controls and of the a.c. system (in particular the system damping and effect of generator control). It appears that a combination of static and synchronous compensators can be the best solution for many applications. The short circuit capacity of the synchronous condenser, coupled with the speed of response of the SVS, should provide better overall voltage control and dynamic performance. This subject has been discussed in Section 5.2.2

8.8.2 Surge arresters
Some early h.v.d.c. schemes used protective gaps but the development of self-sealing surge arresters has changed the situation. Arresters with series gaps and silicon carbide valve elements having non-linear resistance are extensively used in present h.v.d.c. schemes.

With a.c. waveforms the current passes through zero twice per cycle and therefore the voltage drop on the active (or current limiting) gaps merely limits the follow current and makes re-sealing easier. In the case of d.c., re-sealing can take place only if the active gaps build up a voltage at least as high as that against which the diverter has to re-seal.

There is another important difference in the behaviour of a.c. and d.c. surge diverters. Due to the parallel connections, the a.c. system has a low impedance and the follow current in an a.c. surge diverter, even with strong arc-suppression characteristics, has little effect on the a.c. system voltage against which the diverter has to re-seal. On the other hand, d.c. schemes have large smoothing reactors and other inductive components, i.e. the diverter has to suppress the direct current in a highly inductive circuit. As a result, unless the suppression is done in a controlled way, the surge diverter might cause further overvoltages.

A new type of arrester, using zinc-oxide material,[13] has recently appeared which has a more pronounced non-linear resistance characteristic than the silicon carbide; voltage ratings of 588 kV are already commercially available. The main advantages of the new type are its high discharge capability and the lack of gap spark-over transient; the latter property is particularly important in the protection of thyristor valves.

The zinc-oxide arrester voltage characteristic has a very definite "knee" and is extremely flat; hence the arrester will not permit the voltage to rise without shunting a substantial current to ground. Furthermore, in the case of a switching surge, all arresters connected to a busbar share the discharge duty; they draw an increased current out of the overvoltage source and hence contribute to the damping effect.

With gapless metal-oxide arresters, the arrester elements are continuously subjected to the normal operating voltage of the a.c.–d.c. system. The number of series connected elements are selected so that only a very low current flows under normal applied voltage.

8.8.3 Application of surge arresters.

As a general principle, a surge arrester must be set to discharge, or divert, overvoltages higher than the highest normal operating voltage and lower than the breakdown voltage of the insulation under protection.

Most overvoltages within a convertor station are of the switching surge type. While lightning surges caused by thunder storms do not enter the convertor bridges, the valves are exposed to lightning surge type stresses during ground faults within the station. It is thus necessary to test the h.v.d.c. equipment with some standardised voltage waveforms similarly to h.v.a.c. equipment.

Three examples of surge arrester protection in recent schemes are now discussed:

(a) Square Butte: Figure 8.15 illustrates the insulation levels and protective margins for the Square Butte convertors.[14] In this scheme the d.c. arresters are zinc-oxide of early design and thus include a series gap.

D.C. arresters are applied on the line, across the bridges, on the neutral and on the 125 kV bus. Conventional arresters are used across the smoothing reactor, on the transformer primary and across each valve. In addition to valve arresters, each thyristor level is equipped with forward overvoltage protection. The valve arrester characteristics are selected to coordinate with the valve

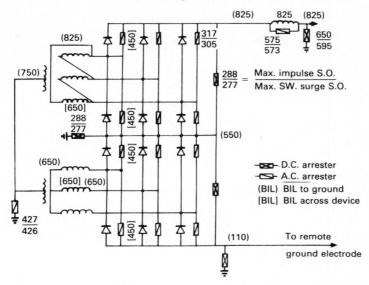

Fig. 8.15 *Square Butte insulation coordination* (© 1980 IEEE)

inverse voltage rating and the operating voltage of the valve forward protection circuits. In this manner the valve is transiently protected against overvoltages whether due to actual system disturbances or due to potential control problems.

It is interesting to note that phase to phase arresters have been omitted in this scheme because the phase to phase insulation is sufficiently high to be protected by the arresters across the valves (two of which are connected between the transformer phases).

(b) Cross-Channel 2000 MW scheme:[15] In this example the emphasis is on the protection and insulation levels adopted for the a.c. and d.c. systems equipment.

Figure 8.16 shows the insulation coordination of the a.c. system equipment. Surge arresters are connected on the 400 kV system from phase to earth, their main function being the limitation of the maximum energy stored in the filter capacitors; this in turn reduces the energy absorption duty imposed on other surge arresters.

Regarding the filter components, during transient conditions the prospective voltage across these components may be even higher than the phase to earth voltage. The insulation level of the resistors and reactors can thus be substantially reduced by using a surge arrester in parallel with these components.

The surge arresters of the d.c. convertor equipment are illustrated in Fig. 8.17. The arrester across the thyristor valve is determined mainly by considerations of maximum continuous operating voltage. For an economic

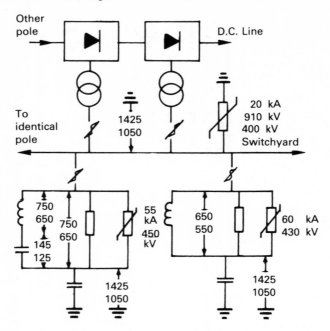

Fig. 8.16 *A.C. equipment insulation coordination–arrester protection levels at 8/20 µs and current specified*

valve design the protection level of the valve arresters should be kept as low as possible. In addition to the discharge energy present in the d.c. system, valve arresters may be exposed to severe discharge duty during fault recovery in the a.c. system; as a result, the arresters incorporate several parallel columns of zinc-oxide blocks.

The d.c. cable arrester is mainly determined by considerations of required protection level. In the absence of overhead lines, fast transient overvoltages of significant amplitude will only be caused by flashover to earth.

(c) *British-Columbia Hydro (Stage IV) 140 kV valves*:[16] Figure 8.18 graphically illustrates the valve insulation coordination and clearly indicates that the thyristor characteristics are far in excess of the arrester characteristics. The central column quantities are in per unit (referred to the rate d.c. voltage).

The commutation transient peak at 90° firing-delay must be less than the minimum sparkover of the arrester; moreover sufficient margin must be allowed for the commutation transient during normally expected overvoltages. Excessive reduction of the commutation transient by damping resistor–capacitor circuits is avoided (considering the increased losses), by introducing an

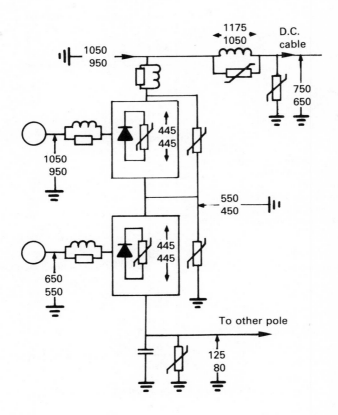

Fig. 8.17 *D.C. equipment insulation coordination—withstand voltages in kV and at 1·2/50 μs and 250/2500 μs*

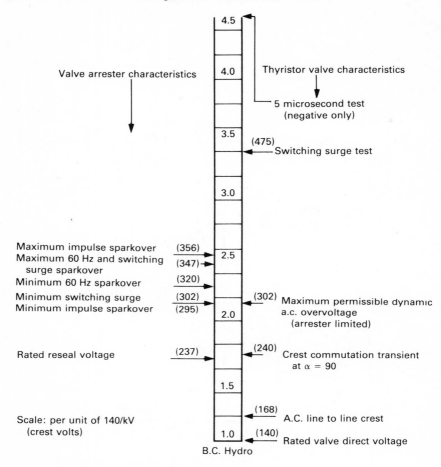

Fig. 8.18 *B.C.–Hydro valve insulation coordination (© 1978 CIGRE)*

inverse-time overvoltage protection scheme which inhibits operation at near 90° firing-delay during excessive a.c. system overvoltage.

8.9 References

1. UHLMANN, E. and FLISBERG, G. (1971): 'H.v.d.c. insulation coordination, Part I: Generation of overvoltages', *Direct Current and Power Electronics*, Vol. 2, No. 1, pp. 8–14.
2. HEISE, W., BURGER, U., KAUFERLE, J., and POVH, D. (1974): 'The Cabora–Bassa D.C. transmission system: overvoltage protection and insulation coordination', *IEEE PES Winter Meeting*, Paper T74 050-1, New York.
3. BOWLES, J. P. (1974): 'Overvoltages in h.v.d.c. transmission systems caused by transformer magnetising inrush currents', *Trans. IEEE*, Vol. PAS-93, pp. 487–493.

4 GIESNER, D. B. and ARRILLAGA, J. (1972): 'Behaviour of h.v.d.c. links under unbalanced a.c. fault conditions', *Proc. IEE*, Vol. 119, No. 2, pp. 209–215.
5 RAYNHAM, E. F. and GOOSEN, P. V. (1981): 'Apollo inverter station h.v.d.c. operating experience', presented to *CIGRE Comittee 14*, Rio de Janeiro.
6 UHLMANN, E. and FLISBERG, G. (1971): 'H.v.d.c. insulation coordination — Part 2: Distribution of overvoltages', *Direct Current and Power Electronics*, Vol. 2, No. 3, pp. 104–111.
7 CLERICI, A. (1973): 'Transient overvoltages caused by earth fault on bipolar d.c. lines', *IEE Conference 107 on High Voltage D.C. and/or A.C. Power Transmission*, London, pp. 196–200.
8 BREUER, G. D., CSUROS, L., HUGUM, R. W., KAUFERLE, J., POVH, D., and SCHEI, A. (1972): 'H.v.d.c. surge diverters and their application for overvoltage protection on h.v.d.c. schemes', *CIGRE Conference Paper 33-14*.
9 ARRILLAGA, J. and EL-BATAL, S. (1973): 'Lightning surge distribution in h.v.d.c. convertors', *Proc. IEE*, Vol. 120, No. 5, pp. 595–600.
10 ARRILLAGA, J. and EL-BATAL, S. (1972): 'Internal oscillations in multibridge h.v.d.c. convertors', *Proc. IEE*, Vol. 119, No. 9, pp. 1351–1359.
11 ARRILLAGA, J. and EL-BATAL, S. (1973): 'Arc-quenching transients in h.v.d.c. convertors', *Proc. IEE*, Vol. 120, No. 11, pp. 1347–1402.
12 DISEKO, N. L., WOODHOUSE, M. L., THANAWALA, H. L., ANDERSEN, B. R., CRAWSHAW, A. M., and ROWE, J. E. (1981): 'Application of a digital computer program to transient analysis and design of h.v.d.c. and a.c. thyristor valves', *IEE Publication 205 on Thyristor and Variable Static Equipment for A.C. and D.C. Transmission*, London, pp. 167–170.
13 KREGGE, J. S. and SAKSHANG, E. C. (1980): 'Zinc oxide arrester experience and application at h.v.d.c. stations', *IEEE Conference on Overvoltages and Compensation on Integrated A.C.–D.C. Systems*, Winnipeg, pp. 65–69.
14 BAHRMAN, M. P. (1980): 'Overvoltage and VAR compensation on the Square Butte h.v.d.c. system', *IEEE Conference on Overvoltages and Compensation on Integrated A.C.–D.C. Systems*, Winnipeg.
15 ANDERSEN, B. R., DISEKO, N. L., and ROBINSON, A. A. (1981): 'Insulation coordination for the U.K. terminal of the 2000 MW h.v.d.c. cross-channel scheme', *IEE Publication 205* (as in reference 12), pp. 199–203.
16 DEMAREST, D. M. and STAIRS, C. M. (1978): 'Solid state valve test procedures and field experience correlation', *CIGRE, Paper 14-12*, Paris.

Chapter 9
D.c. versus a.c. transmission

9.1 General considerations

High voltage transmission serves a dual purpose, that is:

(*a*) System interconnection;
(*b*) Bulk energy transfer.

With reference to system interconnection, the need to operate the whole system in perfect synchronism often prevents the transfer of power by alternating current; typical problem cases have been mentioned in the introductory chapter. In such cases there is no practical alternative to the use of h.v.d.c.

As far as bulk energy transfer is concerned there are various alternatives, not all of them involving electric power transmission, and an accurate economic assessment is essential in each case. Whenever the transmission distance is sufficiently large, and restricting the choice to the electrical alternatives, the case for h.v.d.c. transmission (shown schematically in Fig. 9.1) is well established in spite of the relatively high cost of the dual conversion required.

In less obvious decisions, the accounting procedures used in the economic comparison must include the cost of lines, terminals, any special apparatus needed for voltage support (see Fig. 9.2), short-circuit limitation, etc. The energy lost and the plant needed to supply it must also be capitalised.

However the basic costs alone are not decisive and allowance must be made for other considerations such as:

(*a*) Operational reliability, flexibility and performance during disturbances. The consequences of shutdowns due to maintenance and forced outages.
(*b*) Maximum loading capability as well as the continuous and short-time overload.
(*c*) Transmission system development and the possibility of a staged installation programme.

D.c. versus a.c. transmission

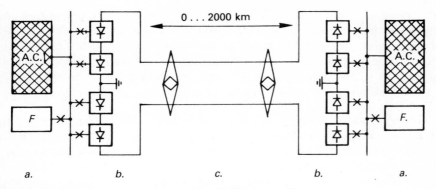

Fig. 9.1 Typical d.c. transmission system (ASEA Journal)
 a = A.C. system
 b = Convertor station
 c = D.C. line
 F = Filter

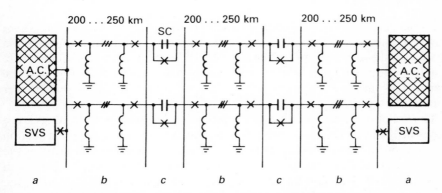

Fig. 9.2 Typical long-distance a.c. transmission system (ASEA Journal)
 a = A.C. system
 b = A.C. transmission line
 c = Substation
 SVS = Static var source
 SC = Series capacitors

Moreover to try and achieve a meaningful and generally applicable comparison in marginal cases is a very difficult task. Among the factors responsible for the complexity of a generalised theory are:

(*a*) The wide range of practical situations involving differing conditions among countries, e.g. overhead line costs vary from country to country by a factor as high as 2·5, while the cost of the convertor terminals varies very little.
(*b*) The lack of technical comparability, given the rather different degrees of freedom of a.c. and a.c./d.c. power systems.

208 D.c. versus a.c. transmission

(c) The need to consider the long-term effects on overall system design and cost when choosing among alternative plans for system development.

(d) The rapid strides being made in the technology of both a.c. and d.c. transmission. In this respect, transmission line costs have experienced a large increase in recent years. As the line cost is relatively lower in the case of h.v.d.c. transmission this effect has affected more the a.c. alternative.

Thus a precise economic determination is only possible in terms of a specific situation, taking into account the long-range development of the system, its probable future pattern of load growth and generation resources, and the many other factors affecting the planning of power systems.

The following sections provide a brief comparison of the a.c. and d.c. technologies with reference to the two basic purposes of power transmission, i.e. Bulk Energy Transfer and System Interconnection.

9.2. Bulk energy transfer

The escalating costs of bulk energy transfer in the first half of the century kept alive the memory of the initial supremacy of direct current as a transmission channel and encouraged its revival. However, the fact that transmission by d.c. requires a less expensive line or cable for the same power capacity has to be weighed against the high cost of the a.c. to d.c. and d.c. to a.c. power conversion terminals.

The use of standardised break-even distances for various power ratings is not necessarily the best criterion for economic comparisons as explained in Section 9.1. A typical example where distance was not the main deciding factor is the case of two Canadian schemes, the Churchill River and the Nelson River, both of which were given consideration at about the same time. Surprisingly, the longer scheme (Churchill) went a.c. and the shorter (Nelson) went d.c. Nevertheless distance is a primary consideration in bulk energy transfer and its influence on the characteristics of a.c. and d.c. power transmission systems is discussed in the following section.

9.2.1 A comparison of A.C. and D.C. transmission characteristics

Overhead transmission: With reference to Fig. 9.3. the power transfer in a transmission line is determined approximately by the expression

$$P = \frac{V_S V_R}{X_{SR}} \sin \theta_{SR},$$

where V represents rms fundamental frequency voltage, X the series reactance, θ the relative voltage phase shift (or load angle) and suffixes S, R indicate sending and receiving ends respectively.

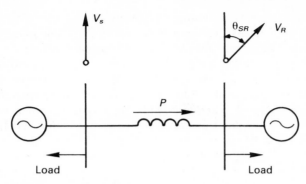

Fig. 9.3 *Power flow in an a.c. system*

Power flow disturbances are quickly reflected in load angle oscillations and therefore, for reasons of stability, the load angle is kept at relatively low values under normal operating conditions (about 30°). For a line loaded with the natural (or surge) impedance this angle puts a limit to the maximum series reactance (0·5 per unit) and therefore to the transmission distance. Beyond this point, to avoid instability due to faults on the line, it is normal practice to provide parallel lines. With very high voltage long distance transmission (Fig. 9.2) it is also usual to include switching stations with series capacitors and other means of voltage support such as synchronous condensers or static compensators.

Moreover the a.c. line conductor has to be designed to transmit the charging current (or capacitive reactive power) under light load conditions. This effect causes overvoltages and requires shunt reactor compensation.

Since under steady state conditions, inductance and capacitance have no effect on a d.c. line, the above difficulties do not arise with h.v.d.c. transmission and there is no need for intermediate switching stations.

On the other hand the operation of a convertor, whether as a rectifier or as an invertor involves a consumption of reactive power. Since the reactive power need is substantial, it is usually supplied as near as possible to the convertor station, partly by the capacitance of the a.c. filters and often by additional shunt capacitors. With weak a.c. systems, the a.c. voltage regulation with varying load conditions may demand the use of synchronous condensers or static compensators.

With d.c. no stability problems occur, because the a.c. systems are decoupled and the power flow can be freely and rapidly adjusted by convertor control. On the other hand overloading is more restricted in d.c. transmission. The silicon controlled rectifier has a very small thermal capacity and thus the modern convertor valve is only designed to handle temporary overcurrents under fault conditions or to damp a.c. system oscillations. If long term overload capability is desired, this can be achieved by appropriate overrating of the thyristors and permitting higher temperatures at the convertor plant.

Cable transmission: High voltage transmission by cable is rarely used because of the higher cost and longer repair times. Their place is normally restricted to underwater crossings and infeed to urban centres.

The high voltage cables have a low series inductance and large shunt capacitance. Moreover their loading, due to the lower surge impedance and thermal limitations is usually below 0·3 times the surge impedance level. Therefore, high charging reactive powers are required, which considerably limit the length of a.c. cable transmission. For instance, at 50 Hz the charging current varies typically from 5·5 A/km for a 132 kV cable to about 15 A/km for a 380 kV cable. With a 4·52 cm², 380 kV cable of 600 A thermal limit, the charging current for a 40 km length equals the thermal limit and no useful load can therefore be carried. Similarly a 2·58 cm², 450 A, 132 kV cable has a critical length of about 80 km.

These critical lengths may be extended by inserting shunt reactors. Even with a 100 per cent compensation by means of two reactors, one at each end, the power transmission capacity is only 86·6 per cent at critical length and reduces to zero at twice the critical length. With two intermediate reactors, each providing 100 per cent compensation, dividing the line in three equal parts, the critical length will increase to three times. Moreover, intermediate compensation is impractical in the case of underwater links.

9.2.2 Power-carrying capability of a.c. and d.c. lines[1]

(*a*) If for a given insulation length, the ratio of continuous-working withstand voltages is

$$k = \frac{\text{d.c. withstand voltage}}{\text{(rms) a.c. withstand voltage}}, \tag{9.1}$$

various experiments on outdoor d.c. overhead-line insulators have demonstrated that due to unfavourable effects there is some precipitation of pollution on one end of the insulators and a safe factor under such conditions is $k = 1$. However if an overhead line is passing through a reasonably clean area, k may be as high as $\sqrt{2}$, corresponding to the peak value of rms alternating voltage. For cables however k equals at least 2 and here the prospect for d.c. is obviously very encouraging.

(*b*) As discussed in Chapter 8 a transmission line has to be insulated for overvoltages expected during faults, switching operations, etc. A.C. transmission lines are normally insulated against overvoltages of more than 4 times the normal rms voltage; this insulation requirement can be met by insulation corresponding to an a.c. voltage of 2·5 to 3 times the normal rated voltage.

$$k_1 = \frac{\text{a.c. insulation level}}{\text{rated a.c. voltage } (E_p)} = 2\cdot5, \quad \text{say, for a.c.} \tag{9.2}$$

On the other hand with suitable convertor control the corresponding h.v.d.c. transmission ratio, i.e.

$$k_2 = \frac{\text{d.c. insulation level}}{\text{rated d.c. voltage } (V_d)} \tag{9.3}$$

need only be 1·7.

Thus for a d.c. pole to earth voltage V_d and a.c. phase to earth voltage E_p the following relations exist:

$$\text{insulation ratio} = \frac{\text{insulation length required for each a.c. phase}}{\text{insulation length required for each d.c. pole}}$$

$$= \left(\frac{\text{a.c. insulation level}}{\text{a.c. withstand level}}\right) \bigg/ \left(\frac{\text{d.c. insulation level}}{\text{d.c. withstand level}}\right)$$

and substituting eqns. (9.1) (9.2) and (9.3)

$$\text{insulation ratio} = \frac{k \cdot k_1}{k_2} \cdot \frac{E_p}{V_d}. \tag{9.4}$$

(c) Consider a new d.c. transmission system to compare with a three phase a.c. system transmitting the same power and having the same percentage losses and the same size of conductor. The d.c. system is considered to have two conductors at plus and minus V_d to earth.

Power in the a.c. system: $3E_p I_L$ (assuming that $\cos \phi = 1$).

Power in the d.c. system: $2I_d V_d$,
          a.c. losses: $3I_L^2 R$,
          d.c. losses: $2I_d^2 R$.

Equating line losses,

$$3I_L^2 R = 2I_d^2 R \tag{9.5}$$

or

$$I_d = (\sqrt{3}/\sqrt{2}) I_L. \tag{9.6}$$

Equating powers,

$$3E_p I_L = 2I_d V_d \tag{9.7}$$

or

$$V_d = (\sqrt{3}/\sqrt{2}) E_p \tag{9.8}$$

and substituting eqn. (9.8) in eqn. (9.4):

insulation ratio $= (kk_1/k_2)(\sqrt{2}/\sqrt{3})$. (9.9)

For the values of k, k_1 and k_2 recommended above, the above ratio is equal to 1·2 for overhead lines and 2·4 for cables.

D.C. transmission capacity of an existing three-phase double-circuit a.c. line: The a.c. line can be converted to three d.c. circuits, each having two conductors at $\pm V_d$ to earth respectively, thus:

Power transmitted by a.c. $P_a = 6E_p I_L$, (9.10)

Power transmitted by d.c. $P_d = 6V_d I_d$. (9.11)

On the basis of equal current and insulation

$$I_L = I_d,$$ (9.12)

$$V_d = (kk_1/k_2)E_p \quad \text{(derived from eqn. (9.4) equated to 1).}$$ (9.13)

The power ratio is therefore

$$\frac{P_d}{P_a} = \frac{V_d}{E_p} = (kk_1)/k_2,$$ (9.14)

and since the actual losses are the same, the percentage power loss ratio will be the inverse of eqn. (9.14). Thus, for the same values of k, k_1 and k_2 as in (*a*) above, the power transmitted by overhead lines can be increased to 147 per cent, with the percentage line losses reduced to 68 per cent and corresponding figures for cables are 294 per cent and 34 per cent respectively.

9.2.3 Equivalent reliability criterion

Before any final cost comparison can be made it is essential to perform reliability studies of the a.c. and d.c. transmission alternatives to ensure that they are reasonably equivalent in this respect.

The reliability of a two-pole d.c. transmission line is often compared with that of a double three-phase a.c. line. However, the validity of such an assumption is questionable if a tower fault, rather than an insulator fault, needs to be considered. A realistic comparison should include both the probability of occurrence of the two types of fault and their associated power transmission loss.

In a recent reliability study carried out to assess the transfer of some 4760 MW in British Columbia[2] over a distance of about 1000 km a minimum of two parallel d.c. transmission lines (as in the a.c. case) was used as a basis for cost comparisons. The compared alternatives were:

a two-line 765 kV series compensated scheme;
a two bi-pole $\pm$ 600 kV d.c. scheme.

Reliability evaluations were performed in two stages:

(a) A Monte Carlo simulation of the radial transmission alternatives, which provided figures of availability of different power transfer levels and frequency of departures from these levels. The studies used information obtained from manufacturers of h.v.d.c. equipment, which included their experience with plant availability, overload capabilities and forced outage rates for each h.v.d.c. component.

(b) A loss of load probability evaluation for the whole British Columbia Hydro system using the information from (a) as input data. An initial comparison without major spares produced very unfavourable relative availability for the d.c. solution. The addition of a spare smoothing reactor and a spare convertor transformer at each terminal, provided similar availability for the two alternatives which were used for the final economic comparison.

9.2.4 Effect of losses and discount rates

The a.c. and d.c. transmission alternatives are costed using a cash flow and present worth analysis; the present worth of the estimated losses should be included.

In the B.C. Hydro study discussed in the previous section[2] the components costs received from the h.v.d.c. manufacturers was carefully analysed, to remove inconsistencies, and often averaged.

It is interesting to consider the large effect that varying the value of losses has on the economic comparison. This is illustrated in Fig. 9.4 for various alternatives in the B.C. Hydro scheme. The d.c. alternative produced the lowest capital cost for low values of the losses. However, the two line 765 kV scheme became cheaper when losses were evaluated at about $6·4 × 10^{-3} per kW-h. It should be pointed out that this situation occurs because of the lower d.c. voltage level used (i.e. 600 kV). With a ± 765 kV d.c. line the relative cost of the d.c. alternative would reduce as the value of losses increased.

Some allowance is made in the economic analysis for the difference between cost of money and inflation. In the above example, such difference was considered equivalent to a discount rate of 3 per cent. The effect of varying the discount rate is illustrated in Fig. 9.5.

9.2.5 Other considerations

Earth return capability: If the circumstances are right, earth may be used as one of the conductors, making it possible to transmit power by one conductor only, with substantial saving in capital cost. For example the Gotland scheme in Sweden consists of only one submarine cable at negative polarity, sea being used as the other conductor. Use of earth, which is a substantially low resistance path, also results in a considerable reduction of transmission losses. No use can ever be made of earth with a.c. systems, because of the associated inductive effects, excepting when it is absolutely necessary to use the rails in electrified railway and even in that case many precautions have to be taken.

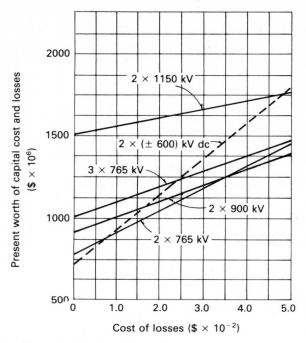

Fig. 9.4 *Cost comparison with varying losses*

The use of earth as a d.c. conductor has difficulties as well, due to possible interference with communication and railway signalling circuits, corrosion of pipes and cable sheaths etc., but in many cases these difficulties do not arise or can be overcome without too much expense.

In many cases only a temporary use of earth may be permissible so that a system with two conductors at $\pm V_d$ is adopted. In such a case, if a fault results in a loss of one conductor, 50 per cent of the rated power can still be supplied through the other conductor and earth. Compared to this a fault on any one conductor of a three-phase a.c. line results in a complete shut-down of transmission.

Favourable routes and right-of-way: Since the relative transmission cost d.c. is lower than a.c., particularly by cable, and as there are no technical limitations with d.c. regarding the length of transmission, it is possible to consider alternative and more favourable routes, from the point of view of way-leaves or security from faults, etc.

The scope of d.c. for transmission of power by cable into or near big cities is enhanced due to its much higher circuit-loading capacity, compared to a.c. cables. Many city centres are now so congested that it is physically impossible to find room for another cable duct or trench, and possibility exists of

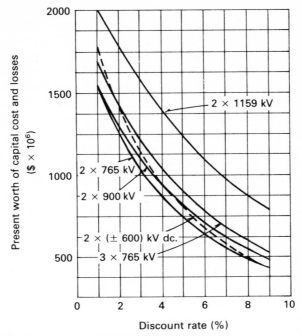

Fig. 9.5 *Cost comparison with varying discount rate*

converting present a.c. cable routes into d.c. cable routes (perhaps by using the existing cables) thereby greatly increasing the circuit capacity; this possibility is discussed further in Chapter 10.

Phased system development: A d.c. terminal consists normally of several convertors connected in series on the d.c. side. The d.c. line voltage can, if desired, be raised gradually by the installation of an increased number of convertors and the power capacity of the d.c. link will be increased accordingly. Both the line and the terminal can in the beginning have a single-pole arrangement using earth return. The capacity can be doubled later by adding another pole.

9.2.6 Infeeds at lower voltage levels

When planning bulk power infeed into load centres the use of underground cable and d.c. transmission will often be favoured. In such cases the choice of the a.c. voltage level at the point of d.c./a.c. power inversion is an important economic parameter.

Large urban areas are normally supplied from a primary high voltage a.c. system (say 500 kV) and one or more levels of secondary distribution systems (of typically 132 kV). As load increases, each of the distribution systems will need reinforcing and eventually, as they reach their specified maximum fault level, new distribution supply points will be required (not normally interconnected). To provide the expanding load new points of generation have to be developed, connected to the primary system; when the fully interconnected primary system reaches its maximum specified fault level it has to be sectionalised. At this point in time a new primary system, at a higher voltage, is normally introduced and to which the new generation is connected.

An alternative system development, which could postpone or eliminate the introduction of a new primary system, consists of h.v.d.c. power injections from the new generation points to appropriate a.c. system locations. The contribution of h.v.d.c. convertor stations to the a.c. network short circuit capacity is negligible as long as no rotating synchronous compensation is required; filter circuit capacitors and capacitor banks have very short time constants during discharge and do not really contribute to the short-circuit power rating.

The resultant effect of such asynchronous power injection on the overall system is illustrated in Fig. 9.6. If asynchronous connected generation were connected at point A, the corresponding amount of reinforcement would be avoided in the primary transmission network. If it were connected at $D1$ and $D2$, reinforcement of both primary and secondary networks would be retarded, resulting in greater gain.

Some hypothetical[3] and realistic[4] supply systems have been investigated which show the economic advantage of h.v.d.c. power injection at the lowest practicable network system level of the metropolitan load area.

9.2.7 Environmental effects[5]

The environmental effects of overhead transmission lines are causing more concern in recent years. Apart from the visual impact, these are mainly related to corona and electric field phenomena, the principal effects being:

(a) audible noise;
(b) electric field;
(c) radio and television interference;
(d) ion production.

(a) Audible noise occurs mainly during wet weather. It is a major design parameter for a.c. transmission and influences the line cost. There is no special noise problem in d.c. transmission because the static field ion production alleviates any surface discontinuities which may appear both under wet and dry weather conditions.

(b) The maximum electric field in the ROW of conductors and at the edge of the ROW are major design parameters in the selection of circuit configuration (i.e. horizontal or delta). The analysis presents no difficulty in the a.c. case, but it is difficult in d.c. lines because in this case space charge effects contribute to

D.c. versus a.c. transmission 217

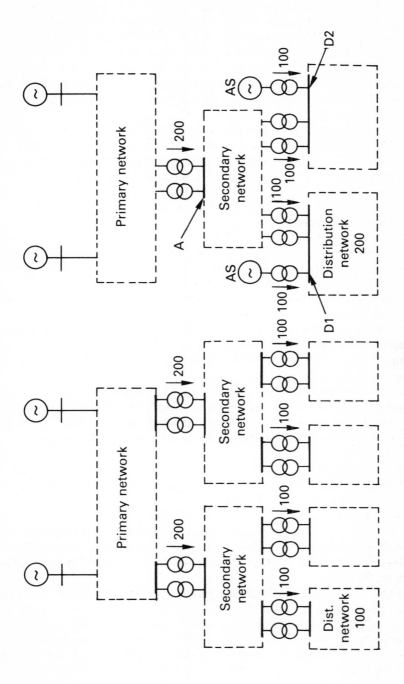

Fig. 9.6 (a) General pattern of development with synchronous sets
(b) Pattern of development with asynchronous and synchronous sets

the earth field. On the other hand the electric field problem is less severe in d.c. because of the lack of steady state displacement current; thus h.v.d.c. lines require much less ROW than the horizontal a.c. configuration and less height than the a.c. delta configuration (i.e. less right of way than h.v.a.c. lines) of comparable rating.

(c) Radio and television interference again constitutes an important design parameter with a.c. while posing no special problem with d.c. transmission design.

(d) Ion production is not a problem with a.c., where any corona activity is essentially contained within one foot of the conductor surface. On the other hand this is a subject still under investigation in the case of h.v.d.c.; line designs are being optimised for low level earth ion current.

9.3 System interconnection

The interconnection of individual power systems brings about considerable advantages such as:[6]

(a) Reduction of installed generating capacity since interconnections permit a lower level of overall reserve power.
(b) Optimal scheduling of hydro-power and thermal power stations, particularly during seasonal fluctuations of water supply.
(c) Improvement of reliability of the interconnected system if reserve capacity remains unchanged.
(d) Generation scheduling may make use of the larger and therefore more economical units. This is particularly important for the smaller networks.
(e) Reduction of peak load of interconnected network due to possible differences in daily, monthly or yearly load cycles of the individual systems or parts of the system.

Moreover the use of a.c. tie-lines is often unsuitable for system interconnections.

(a) Very often the economic power ratings of such interconnections are small in relation to the installed capacity of the systems to be interconnected. In such cases an a.c. tie-line may not be able to cope with the power flow and stability control problems.

An alternative d.c. interconnection provides a fast and flexible power flow control, regardless of the conditions in the a.c. systems, and can provide stability improvement for the two interconnected systems. A typical h.v.d.c. application to this problem is the David Hamil back-to-back interconnection between the eastern and western system at Nebraska (U.S.A), a monopole 100 MW 50 kV link commissioned in 1977.

(b) A.C. interconnections always result in a reduction of the overall system impedance and hence in increases of the short-circuit levels; these may exceed the capability of the existing circuit breakers or cause unacceptable electrical and mechanical stresses on the system equipment. A scheme in which the fault level played an important part in favour of the d.c. solution is the Kingsworth–Beddington–Willesden underground interconnection (U.K.) already described.

(c) If the systems to be interconnected have different frequencies (normally 50 and 60 Hz) an a.c. tie-line is not possible. Examples of d.c. interconnections for this application are the Sakuma and Shin–Shinano links in Japan and the Acaray project in Paraguay (all of them back-to-back).

(d) Even with network systems of the same nominal frequency but controlled according to different principles an a.c. interconnection is often found uneconomical. Early asynchronous (d.c.) ties of this type are the Cross-Channel link (submarine crossing) and the Eel River already described; more recent systems are the USSR-Finland and the Durnrohy (Austria), both back-to-back.

9.4 References

1 ADAMSON, C. and HINGORANI, N. G. (1960): *High Voltage Direct Current Power Transmission*, Chapter 1, Garraway Ltd., London.
2 HARDY, J. E., TURNER, F. P. P., and ZIMMERMAN, L. A. (1981): 'A.c. or d.c., one utility's approach', *IEE Conference Publication 205, on Thyristor and Variable Equipment for A.C. and D.C. Transmission*, London, pp. 241–246.
3 CASSON, W., LAST, F. H., and HUDDART, K. W. (1966): 'The economics of d.c. transmission applied to an interconnected system', *IEE Conference Publication 22*, Manchester, pp. 75–86.
4 EHMKE, B. and HARDERS, C. F. (1980): 'Planning aspects of h.v.d.c. power transmission into metropolitan load centres', *Symposium sponsored by the Division of Electric Energy Systems USDOE*, Phoenix, Arizona, pp. 63–75.
5 LaFOREST, J. J., LINDH, C. B., and STAMBACH, M. R. (1980): 'Techniques for determining overhead line cost data for comparison of a.c. and d.c. transmission alternatives', *Symposium sponsored by the Division of Electric Energy Systems USDOE*, Phoenix, Arizona, pp. 143–161.
6 HARDERS, C. F. and POVH, D. (1980): 'Interconnection of power systems via h.v.d.c. links', *Symposium sponsored by the Division of Electric Energy Systems USDOE*, Phoenix, Arizona, pp. 51–62.

Chapter 10
Research and development

10.1 Introduction

Considerable research and development work is under way to provide better understanding of the performance of h.v.d.c. links, to achieve more efficient and economic designs of the thyristor valves and related equipment and to justify the use of alternative a.c./d.c. system configurations.

This chapter describes briefly various h.v.d.c. proposals which may influence the rate of expansion of d.c. in power transmission; these are:

(a) Present state of development of d.c. circuit breakers as well as their requirements for various d.c. system configurations.
(b) Main characteristics of multiterminal h.v.d.c. systems.
(c) Purpose and operation of single generator convertor units.
(d) Use of forced commutation.
(e) Conversion of present a.c. schemes for use with d.c.
(f) Design and applications of compact convertor stations.
(g) Potential of micro-computers and direct digital control.

10.2 D.C. circuit breakers

Although presently existing h.v.d.c. schemes operate perfectly well without the assistance of d.c. circuit breakers, it is obvious that the prospective extension from point-to-point to other d.c. power system configurations can gain versatility and operational flexibility with the use of d.c. circuit breakers.

Unlike alternating current circuits, where the current passes through zero twice per cycle, no such occurrence takes place in d.c. circuits. This lack of current zeros presents a difficult problem to the opening of d.c. circuits.

Because of the large amounts of energy involved in high voltage schemes, the techniques used in low d.c. voltage circuits are not applicable to h.v.d.c. transmission.

Two different philosophies are used in d.c. fault interruption. One relies on the convertor controls to limit the current to a reasonably low level, so that the duty of the circuit breaker is not too severe. The main disadvantage of this philosophy is that the entire system suffers a voltage depression for the complete duration of the disturbance.

The alternative approach is similar to a.c. system fault clearance, i.e. following fault detection, circuit breaker action is ordered as soon as possible to isolate the faulty section, without relying on any control action by the convertors. This second philosophy is much harder on the circuit breaker since the current to be interrupted will be considerably larger; on the other hand the system disturbance will be minimised, i.e. without the need to depress the system voltage.

An example of the first philosophy is illustrated by the basic circuit of Fig. 10.1,[1,2] which relies on the high arc voltage developed in the switching device to create a current zero. The circuit shows two switches in series, i.e. a commutating switch (CS) and an isolating switch (IS). A commutation circuit,

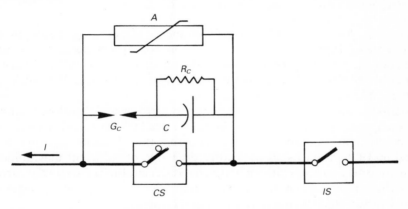

Fig. 10.1 *H.v.d.c. circuit breaker with arc-gap-controlled commutation circuit*

consisting of a capacitor (C) with a discharge resistor (R_c) and a spark gap (G_c), is connected in parallel with CS; a surge arrester (A) is also connected across CS. The arc voltage causes G_c to spark over and thus a sudden current inrush flows into the uncharged capacitor C with a rapid reduction of the current in the arc path of CS. The rapidly rising voltage across C is limited by the surge arrester A which then takes over the full current; from then on the voltage across the breaker is determined by the residual voltage of the surge arrester, which gradually reduces the current to zero.

The isolation switch IS must then be opened before the system voltage is returned to normal.

Various types of h.v.d.c. breaker have been developed using the second philosophy[3]. The circuit of Fig. 10.2 is one of them[4]. Again, it shows a breaker

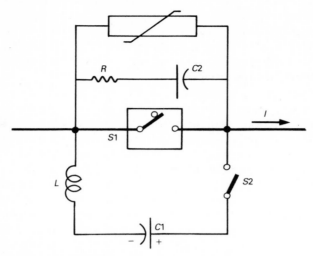

Fig. 10.2 *H.v.d.c. circuit breaker with pre-charged capacitor*

bypassed on the one hand by a capacitor and on the other by a form of d.c. surge diverter. In this case, however, the capacitor is precharged. The breaker used is of the vacuum type because of the outstanding capability of vacuum devices to interrupt high frequency currents and quickly recover their dielectric strength following interruption.

When the contacts of $S1$ separate, the arc is very low and therefore relatively little energy is dissipated in the arc; hence the need for a precharged capacitor. When sufficient separation is achieved across the $S1$ contacts, switch $S2$ is closed and $C1$ produces an oscillatory discharge of sufficient magnitude to create a current zero and thus interrupt the $S1$ circuit. From then on the current flows through the capacitor, until the voltage across is sufficient to trigger the surge diverter which then absorbs most of the energy.

10.2.1 Use of h.v.d.c. circuit breakers in point-to-point interconnections
Two applications have been identified for the use of d.c. circuit breakers to improve the operation and reliability of present a.c.–d.c. schemes, i.e.:

(*a*) metallic return transfer breaker;
(*b*) parallel operation of lines and poles.

(*a*) The metallic return transfer breaker (MRTB) has already been put into operation in several schemes[5]. Its main purpose is to eliminate corrosion in metallic elements close to the d.c. line during monopolar operation. With reference to Fig. 10.3 this is achieved by commutating the current from the ground to a metallic return when a bipolar system is operated in the monopolar mode. A d.c. circuit breaker is included for this purpose in the electrode line.

Research and development

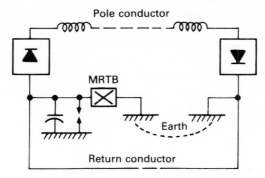

Fig. 10.3 *Configuration using an MRTB circuit breaker*

Although it may appear trivial, the duty of the breaker is considerable as a result of the energy stored in the ground loop; for instance the energy requirement for the Pacific Intertie MRTB is of the order of 8·5 MJ. Moreover the duty will be considerably larger if the non linear resistors, normally provided with the MRTB, are to take over the duty of protecting the neutral during d.c. line fault sequences or pole blocking.

(b) If two separate bipoles are operating with the same nominal voltage, it is possible to connect the convertors in parallel at each end and hence transmit power via two h.v.d.c. lines as shown in Fig. 10.4. It is also possible to isolate one of the h.v.d.c. lines during emergencies, such as faults or maintenance, and transmit full power with one of the two h.v.d.c. lines out of service (subject to thermal limitations).

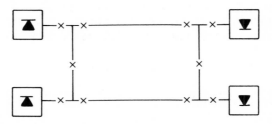

Fig. 10.4 *Parallel convertor/parallel line system*

A report describing the breaker requirements for parallel operation[5] indicates that the use of d.c. breakers, although not essential for this purpose, can result in considerably improved system responses. The energy dissipation varies from a fraction of a MJ, when the breaker is used to direct current into a parallel path, to tens of MJs, in the case of parallel operation with only one d.c. line when the invertor-end breaker is opened following a commutation failure (with no previous control action). In all cases, the energy dissipation could be considerably reduced by increasing the opening time and the nominal voltage of the non linear resistors attached to the switching element.

10.3 Multiterminal d.c. transmission

Although the feasibility of multiterminal d.c. control was demonstrated in the early stages of the h.v.d.c. transmission technology[6,7], there are no true multiterminal h.v.d.c. schemes at present; however two existing schemes, the Nelson River and Kingsnorth, include control characteristics which are virtually multiterminal.

In recent years there has been a revival of interest in the prospective use of multiterminal d.c. schemes[8,9], as the potential benefits of these configurations are recognised. This is mainly due to the operational advantages that have been achieved with two-terminal d.c. systems. Similarly a multiterminal d.c. scheme may provide a considerable improvement in the dynamic performance of an integrated a.c.–d.c. power system, due to the fact that fast power modulation according to selected control strategies can be implemented at more than one terminal.

The performance expected is greatly improved with the availability of fast communication to integrate the overall control strategy; local convertor control can provide the necessary back-up, with some reduction in power transfer, following the loss of communications during contingencies.

Any limitations in the speed of response of proposed multiterminal configurations are mainly due to the characteristics of the main circuit (e.g. smoothing reactors and line capacitance) plus telecommunication delays, and not to the control system itself.

Consideration of multiterminal d.c. schemes must take into account the extra complexity of control, communications and protection systems as well as switching equipment. Particularly important, and difficult to assess, is the question of equalising reliability among alternative multiterminal schemes.

There are three distinct potential applications for h.v.d.c. multiterminal schemes, i.e.:

(*a*) Bulk power transmission;
(*b*) A.C. network interconnection;
(*c*) Reinforcement of an a.c. network.

These can be achieved with various configurations, the basic ones being:[10]

(*a*) parallel or radial tappings (shown in Fig. 10.5);
(*b*) Meshed or ring systems (shown in Fig. 10.6);
(*c*) Series connections (shown in Fig. 10.7).

10.3.1 Technical comparisons
As compared with separate point-to-point interconnections, the parallel multiterminal system of Fig. 10.5 has severe limitations such as:

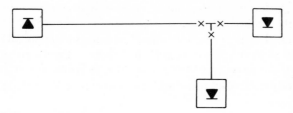

Fig. 10.5 *Line with parallel tap*

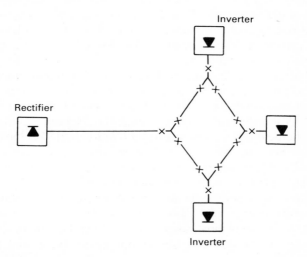

Fig. 10.6 *H.v.d.c. ring system*

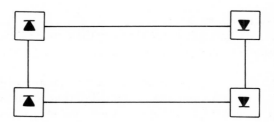

Fig. 10.7 *Series-connected system*

(*a*) The loss of a d.c. line could cause loss of the whole system.
(*b*) A disturbance such as a line fault, or a commutation failure, affects the whole or a large part of the system.
(*c*) Commutation failures at one of the invertors can draw current from the other terminals. If the commutation failure occurs in a station whose rating is small compared with the rating of the other stations recovery will be difficult, if

not impossible, without shutting the whole system down for a short period (or disconnecting the station using a d.c. breaker).
(d) Power reversal at any convertor station requires mechanical switching.
(e) Disconnection of any part of the d.c. line reduces the transmission capacity of the system by at least the MW rating of one convertor station pole and necessitates readjustment of the load flow distribution prior to the disconnection.

On the other hand the parallel system is very flexible for the exchange of power between multiple connection points.

In the case of the ring system (Fig. 10.6) each convertor station is connected to more than one line and therefore the loss of a line has a lesser effect on transmission capability. The other limitations listed for the parallel system are also applicable to the ring system.

The series connection (Fig. 10.7) offers advantages in allowing flexibility of size of tapped station, and for some small ratings of the tap, may be an attractive alternative. On the other hand, the flexibility of power transfer may be inhibited unless a wide on-load tap-change (OLTC) range is incorporated.[7]

With this scheme the main power transfer requirement fixes the transmission line current. Other tapped terminals then operate with this current and their power controllability is achieved by means of d.c. voltage control, which includes OLTC and continuous firing angle control.

Power reversal in the series scheme is possible without mechanical switching and the recovery from system disturbances is similar to the conventional point-to-point links, i.e. in most cases does not affect the remaining system. An overall control system is still necessary for optimised operation of the system.

Insulation co-ordination requires special attention with the series connection.

10.3.2 Economic comparisons

The station costs are basically determined by the power rating, and as each of the configurations is similar in this respect, the terminal costs are expected to be similar.

On the other hand, the level of energy-exchange flexibility required will have an effect on the economic comparison of the different configurations.

Both with parallel and ring configurations an invertor station must be capable of handling extra current from the other terminals during disturbances such as commutation failures; the extra temporary overload capability required for this purpose will of course influence the cost of the terminal.

The terminal losses are likely to be different for the various configurations. The highest are expected with the series connection on account of the need to operate at high firing angles as compared with the parallel and ring systems.

The transmission costs include the cost of the line plus the losses; the specific line costs per unit distance and power generally decrease as the power rating increases. Therefore the transmission line costs of a parallel radial system can be expected to be relatively low; on the other hand the loss of a line might result in a complete system shut-down in this case, which would not normally be acceptable. Hence economy cannot be evaluated without having regard for system security.

With the ring system under normal operation the current will be divided and thus a smaller conductor cross-section can be expected. However, on loss of a conductor the remaining paths must be capable of coping with the overload.

In the series-connection the total current flows through the whole system at all times, and therefore the losses can be expected to be higher than with the radial or meshed systems.

10.3.3 Fault detection

Various principles can be used for the discrimination of an earth fault. The most direct technique is based on continuous monitoring of current differences in different parts of the system. However the need for telecommunications is a drawback and other methods based on the measurement of voltage and rate of change of voltage, possibly combined with current and rate of change of current, are recommended for fast detection of line faults.

However, when voltage and current waves are travelling long distances on overhead lines, or on mixed overhead and cable lines, they get distorted; for this reason it is often difficult to distinguish an earth fault from a commutation failure in less than 20–30 ms.

It should also be noted that with a line fault in a ring system, the voltage and current transients will travel around the system in a complicated way. Therefore it would be difficult to discriminate with certainty the faulty line without the use of information from the other end. This means that telecommunication times have to be included in the time for detection of the line in which the fault has occurred.

10.3.4 Switching requirements

In order to connect and disconnect convertor stations and transmission lines in h.v.d.c. multiterminal systems according to the various alternatives presented in Section 10.3.1 various switches or breakers have to be installed in the stations.

During normal operation the current in the convertor stations, and also in the lines of a parallel system, can be controlled to very small values or to zero and therefore the circuit breaker requirements at normal operation can usually be met by fast acting isolators or breakers designed for a.c. systems. Such is not the case in ring systems as the current in the individual lines cannot be easily controlled to zero without upsetting the total power flow. In this case the d.c. line breaker has to generate a sufficient voltage to commutate the current to a parallel line path.

The switching requirements are of course more stringent during fault clearance.

Let us consider the case of an earth fault occurring between the station breaker (with the station on the invertor mode) and the convertor unit. If the station breaker opens very quickly (say in 2 ms after fault detection), only the porportion of breaking voltage which is in excess of the internal feeding voltage of the system will be effective in breaking the current. As the internal voltage in the system is approximately equal to the rated network voltage before any control action is taken, a fast breaker has to generate a voltage that is appreciably higher than the rated system voltage. Such a breaker will also generate transient overvoltages in the system and therefore overvoltage limiting devices have to be used; these can either be an integral part of the breaker, or separate devices, e.g. line and station arresters.

The energy to be dissipated during the earth fault in a ring system will be much higher than that involved when disconnecting a line in normal operation. It will be at least twice, assuming the same inductance and current and with a breaker voltage that is 60% higher than the rated line voltage. The energy dissipation requirement will be even higher if a lower voltage breaker is used; such a breaker may have been chosen to limit the overvoltage surges at breaker opening during normal operation.

Since the energy is proportional to the square of the current, it is very important to achieve the full breaking voltage in the shortest possible time. Such a breaker is also said to have a current-limiting action, as the fault current is limited by the operation of the breaker.

An alternative to the use of breakers with very fast operating time is to use breakers with moderate operating times, in the order of 30–50 ms. A suitable current control system should be capable of reducing the current in the faulty station below the rated value within 20–40 ms. The dissipated energy in the breaker will then be much less than for the preceding case and it may be acceptable to use breakers with a similar lower breaking voltage, e.g. 60% of the rated line voltage, as in the case of normal operation in ring networks.

The advantage, with breakers operating after the current has been decreased by control action, from the system point of view, is that there is almost no risk of invertor overloading, as the current orders in the rectifier can be changed by telecommunication signals before the system is re-energised. The disadvantage is, of course, that the time during which the power transmission is completely or partly interrupted will be longer.

This example with an earth fault in the station, where the fault detection problem is simple, has demonstrated that overvoltage transients in the system, possible overloading of stations and duration of system shut-down have to be considered when the protection principles and breakers requirements are to be specified. Such requirements must include: maximum dissipated energy, magnitude of breaking voltage, time from tripping order to full breaking voltage and time from current extinction to full voltage withstand capability. Very large systems consisting of many h.v.d.c. stations will probably require fast h.v.d.c. station breakers while breakers with more moderate speed might be sufficient for smaller systems.

10.4 Generator–Rectifier units

In existing h.v.d.c.-connected power stations the generator–transformer units are connected in parallel. An alternative 'unit-type' scheme has been suggested[11], in which each generator operates independently and is directly connected to a convertor transformer unit.

The advantages of the unit connection as compared to conventional rectifier stations are as follows:

(*a*) Since each generator operates independently from the others there are no synchronisation or stability problems.
(*b*) There is no need for special measures to ensure the balancing of reactive power between generators.
(*c*) The a.c. filters can be eliminated and with them the risk of self-excitation of the generators following load rejection. Moreover, any harmonic resonances which would normally occur between the transformers or the generators and the filters are avoided.
(*d*) As a result of the generators segregation the fault level of the a.c. system is substantially reduced with considerable savings in generation, transformer and switching plant. Also the asynchronous nature of the interconnection permits generating at more economical frequencies.

With the use of single generator–convertor units the control of the station can be carried out mainly by the rectifier bridge, resulting in a much simpler generator control, or by the generator, with the possibility of a diode-rectifier convertor. These two possibilities are next discussed in some detail.

10.4.1 *Unit connection using controlled rectifiers*[12,13]
With this scheme the rectifier fast current control determines the transmitted power and provides the necessary valve protection. Only a slow turbine speed control is needed to determine the frequency of the set. Again a slow voltage regulator is sufficient to determine the ideal no-load d.c. transmission voltage.

The operation and protection of this scheme is basically the same as those of present schemes. This system is capable of power reversal and variable speed which might be of interest in pump-storage applications.

If required, several valve groups could be connected in series or parallel on the d.c. side as shown in Fig. 10.8.

The possibility of using a self-excited induction generator together with a controlled rectifier has been investigated, both for long distance d.c.

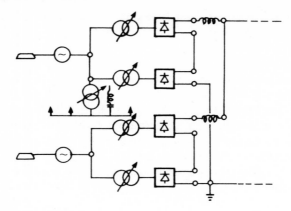

Fig. 10.8 Unit-type scheme

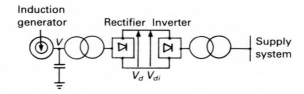

Fig. 10.9 Asynchronous power injection through rectification and inversion

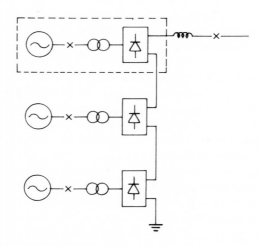

Fig. 10.10 Generator diode-rectifier system

transmission from remote hydro power sources[14], illustrated in Fig. 10.9 and for wind power applications.[15]

10.4.2 Unit connection using diode rectifiers[16]

Figure 10.10 shows a series connection of generator-diode rectifier units. Apart from the need for a faster generator-excitation control the a.c. side of the rectifier is basically the same as in the case of a controlled rectifier unit.

The current harmonic content is substantially smaller with diode operation as a result of the increased commutation overlap and therefore the elimination of a.c. harmonic filters is more justifiable.

In this case it is preferable to use the generator excitation to control a constant d.c. voltage at the sending end. The fast current control is taken over by the invertor whose extinction angle is to be controlled by the slower action of the convertor transformer tap-changer.

In order to prevent frequent commutation failures it is necessary to operate the invertor with higher angles of firing advance. This results in higher costs for the invertor plant (by about 5–10%) and the need for extra reactive power compensation at the invertor end (15–25%).

With the generator-diode rectifier unit most of the conventional h.v.d.c. sending end convertor plant is eliminated. The diode rectifier occupies less space than a corresponding thyristor unit and if it is built, together with the generator and transformer, inside the generator building, the h.v.d.c. terminal is practically eliminated.

The reliability of a diode rectifier should be even better than that of a corresponding thyristor unit, because of its simplicity, without the need for complex grading, control, communication and damping requirements of the controlled rectifier.

For isolated generation (e.g. hydraulic, mine-mouth or nuclear) the diode rectifier scheme appears to offer substantial reductions in h.v.d.c. equipment and provide a more reliable and easily maintained system.

However the practicability of the generator-diode rectifier unit relies on the capability of the generator exciter to provide sufficient protection and control during d.c. line short-circuits. The dynamic response has been tested on a simulator[17] and digital computer[18] but so far no verification has been carried out in a real system.

Upon detection of a d.c. short-circuit (assumed to be achieved in half cycle) the thyristor exciter firings are given the maximum delay. The fault causes an increase in voltage regulation which in conjunction with fast de-excitation, causes the generator terminal voltage to decrease rapidly; this in turn extends the commutation periods and further depresses the voltage. The voltage depression in this case limits the peak currents to under 2 p.u. within two cycles of fault occurrence.

In order to clear the fault quickly and increase system security, the use of a d.c. circuit breaker is recommended. The rating of such breaker, however, can be substantially reduced if it is used in combination with the generator a.c. circuit breaker.

Clearing the fault exclusively by a.c. circuit breakers at the rectifier is not a practical proposition, except for permanent faults, because conventional breakers are too slow in re-close, and because this action does not actually stop the d.c. current immediately; bypass action takes place naturally and the peak d.c. current decays over several seconds.

Another problem is the behaviour due to either momentary commutation failure or temporary a.c. fault at the invertor end. In either case it would take a long time to reduce the resulting overcurrent, then restore it after the fault is removed because of the relatively slow field controllability (as compared with the conventional thyristor control); such time may prove unacceptable for the receiving systems if the proportion of d.c. power is large.

10.5 Forced commutation

With the bridge configuration used in present h.v.d.c. schemes the commutation is only possible when the incoming phase voltage is more positive than that of the outgoing phase; such transfer of current from valve to valve is known as natural commutation. As a result, the operating range of the bridge convertor is confined to quadrants 1 and 2 (shown in Fig. 10.11) and both processes of rectification and inversion consume reactive power in proportion to the power factor angle (refer to eqn. (2.26)).

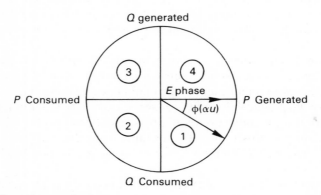

Fig. 10.11 *Operating quadrants*

Commutation in quadrants 3 and 4 can be forced by a temporary reversal of the valve-to-valve voltage during the commutation period; in this way the convertors could be made to operate at a leading power factor. The transfer of current under these conditions is known as forced commutation.

A typical circuit used in forced commutation is illustrated in Fig. 10.12 which includes auxiliary thyristors (A1–A6) and capacitors; the capacitors are

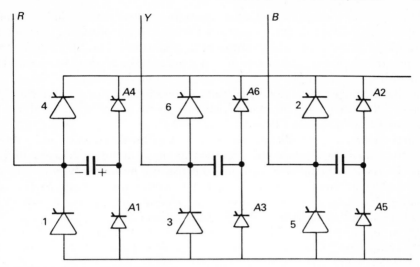

Fig. 10.12 *Forced commutated circuit*

charged, prior to the commutation forcing, with the polarity shown in the diagram (for a commutation from 4 to 6). However the 'capacitive' commutation causes considerable voltage stress on the valves and convertor plant.[19] Thus the economic case for forced commutation is very weak at the moment.

It has been argued[20] that the possibility of valve turn-off can provide fast and secure recovery from commutation failures, a.c. faults, etc.; it can also permit operation with extremely weak a.c. systems. Moreover, the combined use of forced and naturally commutated convertors could provide complete independence of real and reactive power controls, i.e. permitting power flow control through the h.v.d.c. link while maintaining voltage control on the a.c. busbar.

With recent advances in the solid state switching technology[21] the future use of forced commutating techniques for particular applications may not be out of the question.

10.6 Existing a.c. transmission facilities converted for use with d.c.[22]

Normally a.c. transmission lines are not loaded to their maximum thermal rating and the firm power capability is lower in any multi-circuit a.c. arrangement than it could be if the same facilities were used to transmit power by d.c. If any a.c. link is used for d.c., the conductors can form poles of a d.c. system which may be operated independently, if necessary, up to the thermal rating.

The insulation, chosen to satisfy a.c. system requirements, could generally sustain a d.c. voltage to earth equalling the peak value of the a.c. voltage to earth, or exceeding it if heavy atmospheric pollution is not a limiting factor. The possibility of using earth return under outage conditions offers an added attraction.

Similar arrangements could be made for both land and submarine type cables. This principle has already been applied in the case of the d.c. transmission to Vancouver Island.

The conversion capability is not restricted to overhead lines and cables. Capacitor units for series or shunt compensation, if suitably specified initially, could be reused at convertor stations to constitute harmonic filters. Existing a.c. switchgear could also be reapplied at convertor stations to provide isolation facilities and, in the event of schemes with earth return, a.c. switchgear could be used for high-speed changeover duty.

An additional benefit will be a reduction in the fault level.

The major problems expected in the implementation of this method of increasing power transmission capability are:

(*a*) The withdrawal from service during the period of changeover.
(*b*) The need, in many schemes, to provide loads at intermediate points.

Some practical examples illustrating conversion possibilities are shown in Figs. 10.13 and 10.14. The first example illustrates a transmission reinforcement and the second refers to an asynchronous reinforcement of a distribution system in congested areas.

This method is a practical alternative to the addition of further parallel a.c. circuits or the introduction of a higher a.c. voltage. It thereby provides an economic means of overcoming potential amenity problems, which may eventually emerge as the limiting factor of public acceptance of very high voltage overhead transmission.

10.7 Compact convertor stations

Many prospective h.v.d.c. applications involve at least one terminal in the vicinity of metropolitan areas and often the potential convertor sites are subject to air pollution. Such cases can benefit from compacting techniques.

A report of an EPRI sponsored project[23] describes the development of gas-insulated bus systems and compact valves of a so called 'dead-tank' design. SF6 gas-insulated bus is used to interconnect the d.c. potheads, smoothing reactors, valves and convertor transformers. The use of a gas bus in combination with air-insulated valves and conventional valve buildings yields the following benefits:

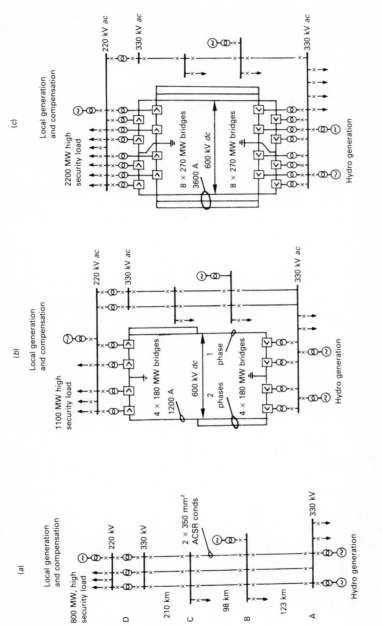

Fig. 10.13 *Power transmission reinforcements*
(a) Existing 330 kV a.c. system
(b) One a.c. circuit converted for d.c.
(c) Two a.c. circuits converted for d.c.

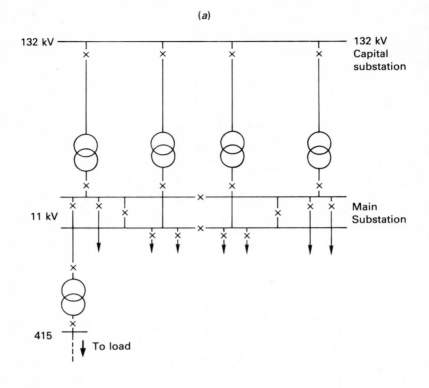
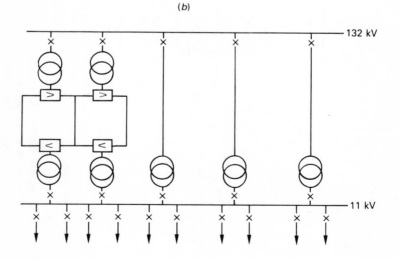

Fig. 10.14 *Distribution system reinforcement*
(a) Conventional a.c. system
(b) Conversion of one feeder for d.c.

(a) minimum valve building size and cost;
(b) avoidance of pollution deposits on exposed portions of a d.c. yard; this would avoid the regular silicon grease applications that are used in some places today to combat pollution problems.

Regarding valve compacting, the primary means of achieving it are the use of gas insulation and liquid cooling; the removal of heat is achieved by the use of liquid freon. The use of a freon cooling system is probably the only realistic approach for dead-tank designs; it may also be an attractive alternative to water for valve-cooling systems, since it is a much superior dielectric as compared to water, and leaks do not cause flashovers.

Clearly, the primary application of the technology is in urban areas. The dead-tank design not only minimises ground space for the terminal, but should also minimise the volume needed for it. This is an important consideration since building costs are a function of volume. It may permit the station to be placed in the basement of a new major downtown office building. If the compact station has limited height above ground it may be possible to be placed under a parking house in a city centre, or even in a residential area.

Other examples of technological progress being made are:

(a) The development of direct, light-fired thyristors, which could greatly reduce the number of other electric components in the valve and increase overall valve reliability and efficiency.
(b) Increases in thyristor voltage rating and protection to reduce the number of series-connected thyristors for the required voltage ratings, reducing valve losses and size. A neutron irradiation technique is already being used for silicon cell doping which increases substantially the thyristor voltage rating.

10.8 Microprocessor-based digital control

Although the gathering of accurate digital information has been advocated for over a decade[24-27], its dependence on hardwired logic has been a factor delaying the implementation of direct digital control in h.v.d.c. transmission. With the appearance of the microprocessor, the situation is likely to change and, although there is still some reluctance to its acceptance in h.v.d.c. transmission as the main control component, there is no question about its role in the processing of information.

H.v.d.c. transmission is ideally suited to digital control[28,29] since control over the convertors is only exercised at certain discrete points in time. Thus the interval between these points is available for the digital calculation of the next firing instant.

The problem arises, however, that the actual time available for the calculation is relatively short. Only thirty of the sixty degrees between firings may be used for calculation purposes. With a 6-pulse convertor operating on a 60 Hz system the time available for control calculation purposes is about 1 ms. Thus if a single microprocessor was used, the basic control algorithm would occupy virtually all the available execution time. This constitutes a serious drawback if other tasks are to be executed in addition to the basic control.

One possible solution to the problem is to increase the speed of the microprocessor. This solution, however, is both expensive and difficult to implement, since the speed of all the peripheral devices must also be increased. The alternative solution is to increase the number of processors and use them to process the individual tasks simultaneously.

The availability of relatively cheap microprocessors, coupled with the development of bus structures capable of supporting multimaster systems, enables the development of a simple but effective h.v.d.c. control and monitoring facility. A basic system, illustrated in Fig. 10.15, consists of several single board computers and data acquisition boards that interface to the convertor model.

The data acquisition boards contain the necessary number of analogue-to-digital convertors to provide information of the commutating voltage waveforms, d.c. current, rate of change of d.c. current, and d.c. voltage signals for use by the control and monitoring systems.

Another important function of these boards is the generation of interrupts based on changes in the valves ON/OFF states and the commutating voltage zero crossings. These interrupts are used to drive the appropriate software angle-measuring routines. They also have an accurate a.c.-period measuring function, to provide the controller with accurate indication of the a.c. system frequency.

The most time-critical software tasks are the control interrupts, which are allocated the highest priority interrupt levels.

Whenever the controller board is not executing an interrupt service routine, it is available to execute other less urgent tasks.

The use of parallel processing provides the basis for many special tasks outside the scope of present analogue-based systems, e.g.:

(a) various types of control can be implemented with the same hardware;
(b) the control mode and the control characteristics can be altered by software to optimize the convertor operation
(c) considerable amount of information can be stored for diagnostic purposes. In particular the facility available to store complete responses digitally, permits observing the effect of changes in the a.c. system or control parameters and directly comparing them with previous trials.

The direct digital control and monitoring facilities provided by the microprocessor-based scheme are illustrated in Fig. 10.16. The graph shows the dynamic variation of extinction angle following a large step change in

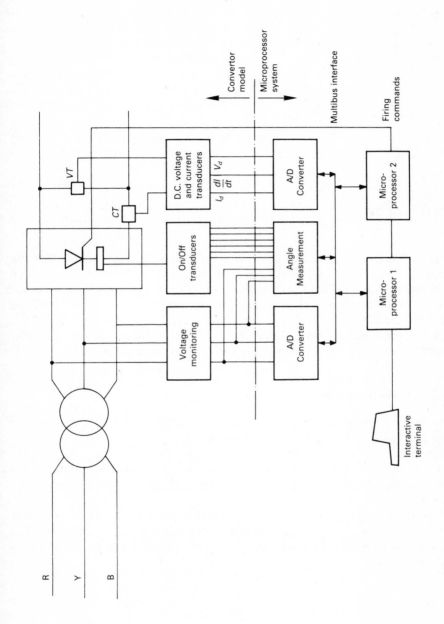

Fig. 10.15 Simplified diagram of a microprocessor-controlled system

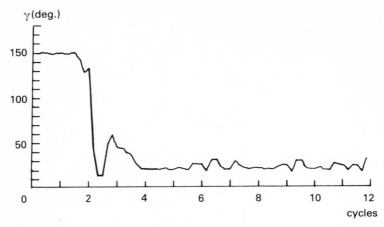

Fig. 10.16 *Display of the extinction angle in an experimental 6-pulse bridge under microprocessor control following a step change in the current setting from 1 to 0.3.*

current order (from 1 to 0.3 p.u.) in a small experimental model. For the purpose of the demonstration the gain of the constant current loop has been made artificially high and the figure shows that the overshoot causes a change from constant current to minimum extinction angle control, with a detailed record of the extinction angle during the transient.

A microprocessor-based system can also be programmed to provide other useful information such as the voltage and current harmonic content.

10.9 General conclusion

Whether the developments described in this chapter are taken up or not, it is now generally accepted that the thyristor valve and its associated h.v.d.c. technology is a success story. It is hoped that the book has been able to demonstrate that h.v.d.c. transmission is already a reliable, efficient and cost-effective alternative to h.v.d.c. for many applications; often regardless of transmission distance.

Much effort is being devoted at present to further research and development in solid-state-based technology, as a result of which it can be expected that h.v.d.c. convertors and systems will play an even greater role in future power systems.

10.10 References

1 KIND, D. *et al.* (1968): 'Circuit breakers for h.v.d.c. transmission', *CIGRE Report 13-08*, Paris.

2 EKSTROM, A. et al. (1976): 'Design and testing of h.v.d.c. circuit breakers', *CIGRE Report 13-06*, Paris.
3 GREENWOOD, A. (1980): 'H.v.d.c. circuit breakers — where do we stand?', *Symposium sponsored by USDOE on Incorporating H.v.d.c. Power Transmission into System Planning*, Phoenix, Arizona, pp. 399–412.
4 GREENWOOD, A. and LEE, T. H. (1972): 'Theory and application of the commutation principle for h.v.d.c. circuit breakers', *Trans. IEEE*, Vol. PAS-91, pp. 1570–1581.
5 BOWLES, J. P. and NILSSON, S. (1980): 'Several possible applications of h.v.d.c. circuit breakers', *Symposium sponsored by USDOE on Incorporating H.v.d.c. Power Transmission into System Planning*, Phoenix, Arizona, pp. 425–444.
6 LAMM, U., UHLMANN, E., and DANFORS, P. (1963): 'Some aspects of tapping of h.v.d.c. transmission systems', *Direct Current*, Vol. 8, pp. 124–129.
7 REEVE, J. and ARRILLAGA, J. (1965): 'Series connection of convertor stations in an h.v.d.c. transmission', *Direct Current*, Vol. 10, pp. 72–78.
8 AINSWORTH, J. D. (1977): 'Multiterminal h.v.d.c. systems', paper presented at *CIGRE Study Committee 14*, Winnipeg.
9 BREUER, G. D. et al. (1980): 'Multiterminal h.v.d.c. networks', *Symposium sponsored by USDOE on Incorporating H.v.d.c. Power Transmission into System Planning*, Phoenix, Arizona, pp. 445–471.
10 KANNGIESSER, K. W., BOWLES, J. P., EKSTROM, A., REEVE, J., and RUMPF, E. (1974): 'H.v.d.c. multiterminal systems', *CIGRE Report 14-08*, Paris.
11 CALVERLEY, T. E., OTTAWAY, C. H., and TUFNELL, H. A. (1973): 'Concepts of a unit generator convertor transmission system', *IEE Conference Publication 107 on High Voltage D.C. and A.C. Transmission*, London, pp. 19–23.
12 KRISHNAYYA, P. C. S. (1973): 'Block and double-block connections for h.v.d.c. power station infeed', *IEEE PES Summer Meeting*, Conference paper C73227–6.
13 ARRILLAGA, J., CAMPOS BARROS, J. G., and AL-KHASHALI, H. J. (1978): 'Dynamic modelling of single generators connected to h.v.d.c. convertors', *Trans. IEEE*, Vol. PAS-97, No. 4, pp. 1018–1029.
14 ARRILLAGA, J. and WATSON, D. B. (1978): 'Static power conversion from self-excited induction generators', *Proc. IEE*, Vol. 125, No. 8, pp. 743–746.
15 WATSON, D. B., ARRILLAGA, J., and DENSEM, T. (1979): 'Controllable d.c. power supply from wind-driven self-excited induction machines', *Proc. IEE*, Vol. 126, No. 12, pp. 1245–1248.
16 BOWLES, J. P. (1978): Discussion of Reference 13, pp. 1027–1028.
17 BOWLES, J. P. (1977): 'H.v.d.c. system developments and concepts — The diode rectifier', *CIGRE Study Committee 14*, Winnipeg.
18 HEFFERNAN, M. D. (1980): 'Analysis of a.c./d.c. system disturbances', Ph.D. Thesis, University of Canterbury, New Zealand.
19 UHLMANN, E. (1975): *Power Transmission by Direct Current*, p. 123, Springer-Verlag, Berlin–Heidelberg.
20 BOWLES, J. P. (1980): 'Alternative techniques and optimisation of voltage and reactive power control at h.v.d.c. convertor stations', *IEEE Conference in Overvoltages and Compensation on Integrated A.C.–D.C. Systems*, Winnipeg, pp. 5–5i.
21 BALIGA, B. J. (1981): 'Switching lots of watts at high speeds', *IEEE SPECTRUM*, Vol. 18, No. 12, pp. 42–47.
22 JONES, K. M. and KENNEDY, M. W. (1973): 'Existing a.c. transmission facilities converted for use with d.c.', *IEE Conference Publication 107 on High Voltage D.C. and/or A.C. Power Transmission*, London, pp. 253–260.
23 FLAIRTY, C., HINGORANI, N. G., LEBOW, M., and NILSSON, S. (1980): 'EPRIs compact, gas-insulated d.c. convertor stations', *Symposium sponsored by USDOE on Incorporating H.v.d.c. Power Transmission into System Planning*, Phoenix, Arizona, pp. 507–518.

24 REEVE, J. (1967): 'Direct digital protection of h.v.d.c. convertors', *Proc. IEE*, Vol. 114, No. 12, pp. 1947–1954.
25 ARRILLAGA, J. and GALANOS, G. (1970): 'Theoretical basis of a digital method of grid control for h.v.d.c. convertors', *Trans. IEEE, Power Apparatus and Systems*, Vol. PAS-89, No. 8, pp. 2049–55.
26 ARRILLAGA, J. and BALDWIN, D. G. (1972): 'Derivation of logical information for the direct digital control of h.v.d.c. convertors', *Trans. IEEE, Power Apparatus and Systems*, Vol. PAS-91, No. 2, pp. 360–367.
27 ARRILLAGA, J. and HISHA, H. 'Fast ON/OFF detection of silicon controlled rectifiers and its influence on convertor controllability', *Trans. IEEE on Industrial Electronics and Control Instrumentation*, Vol. IECI-26, No. 1, pp. 22–26.
28 REEVE, J. and GIESBRECHT, W. J. (1978): 'Evaluation of a microcomputer for h.v.d.c. convertor control', *IEEE PES Summer Meeting*, Paper A78 550-6.
29 SHORE, N. L. and FRERIS, L. L. (1978), 'Minicomputer on-line control of d.c. link convertors', *Proc. IEE*, Vol. 125, No. 3, pp. 215–220.

Subject index

AC-DC system interaction, 99, 121
AC system faults (see Short circuit faults)
AC switchyard, 141, 156
AC transmission, 1, 207, 208
Anode, 4, 16
Anode reactor, 143, 146
Arcback (see also Backfire) 3, 9, 15
Asynchronous interconnection, 1, 6, 9, 216, 219
Audible noise, 216
Average direct voltage, 30, 33

Backfire, 162, 175, 180
  consequential, 175
Blocking of valves, 164, 166, 180, 188
Break-even distance, 208
Bridge convertor, 16
Bridge inverter, 31
Bridge rectifier, 16, 28
Bridges in parallel, 22, 133
Bridges in series, 130, 133, 183
Bypassing of bridges, 133, 164
  use of main valves for, 165
Bypass switch, 146, 165
Bypass valve, 146, 164

Cables, 151, 210
Cathode, 16, 143
C.E.A. control (see Constant extinction angle control)
Charging current, 209, 210
Circuit breakers, HVDC (see DC circuit-breakers)
Commutating voltage, 22, 77, 170
Commutation (see also Overlap angle) 16, 22, 77
Commutation angle (see Overlap angle)

Commutation failure, 159, 170, 177, 180, 188, 225
Commutation reactance, 22, 24, 31, 91, 104
Commutation transients, 184, 202
Computer modelling, 67, 114, 167, 171, 175, 196, 231
Constant current control, 80, 88, 90
Constant current transmission, 88
Constant extinction angle control, 79, 81, 109
Control
  analogue, 77, 87
  characteristics, 89, 93
  constant current (see Constant current control)
  constant extinction angle (see Constant extinction angle control)
  equidistant, 45, 79, 82, 114
  digital, 87, 237
  filters, 79
  frequency, 115
  individual phase-control, 77, 106
  instability of, 81, 87, 105
  multiterminal (see Multiterminal dc)
  power, 91, 96, 116, 118
  predictive, 78, 86, 109
  tap changer (see On-load tap-changing)
Convertor
  back to back, 47, 154, 181, 219
  compact, 234
  bridge circuit, 16, 37
  mercury-arc, 143, 193
  multibridge, 25, 26, 40
  phase-shift, 23, 40, 52
  pulse number, 37, 43, 51
  rating factor, 99, 101

Subject index

six-pulse, 29, 32, 45
twelve-pulse, 41, 46, 65, 86, 127, 194
Current extinction, 188, 190
Current margin, 89
Current setting, 89

DC circuit breakers, 220, 227, 231
DC line, fault, 173, 181
DC link
  bipolar, 123, 129, 138
  construction in stages, 215
  monopolar, 133
DC reactor, 15, 18, 30, 68, 140, 142, 149, 190
DC transmission, advantages, 1, 209, 213, 215, 218
Deblocking, 185
Deionization of fault arc, 174
Delay angle (see also Gate control) 21, 33, 41, 77, 91
Direct voltage (see Average direct voltage)
Displacement factor, 34
Distortion of alternating voltages (see Voltage distortion)
Dynamic compensation, 103, 105

Earth return, 5, 6, 213, 215
Economics, 103, 126, 143, 226
  ac versus dc, 206, 212
Electric field, 216
Electrode, earth, 152
Environmental effects, 216
Extinction angle, 9, 31, 42, 77, 82, 92

Filter
  admittance of, 60
  cost of, 61, 63
  damped, 63, 66, 192
    second-order, 64
    third-order, 64
    type-C, 64
  dc, 67
  double-tuned, 60
  high-pass (see Filter, damped)
  impedance of, 59
  protection of, 182
  quality of, 53, 59, 61, 63, 68
  reactive power of, 53
  self-tuned, 63
  sharpness of tuning, 53
  single-tuned, 59

  size, 53
  tuned, 59
Filter design, 52
Firethrough, 159
Firing angle (see Delay angle)
Firing error, 43
Forced commutation, 232
Fourier analysis, 31, 35, 37, 41, 46
Frequency control, 6, 115
Frequency conversion, 6
Frequency deviation, 62, 118

Gate control, 18
Generators feeding dc link, 229
  unit connection, 229, 331
Grading electrode, 3, 5
Grid control, 2

Harmonic distortion, 45, 167
Harmonic filters (see Filter)
Harmonic impedance, 53, 55
Harmonic injection, 70, 72
Harmonic instability, 99, 106, 112, 114
Harmonics, 35
  ac characteristic, 37
    at no overlap, 41
    at overlap, 42
  direct-voltage, 37
  even, 43, 112, 187
  magnification of, 79, 87, 106
  resonance, 57, 59, 61, 64, 108, 112
  triplen, 43, 46, 64
  uncharacteristic, 42, 115

Instability (see Harmonic instability and power instability)
Insulation co-ordination, 104, 185, 198, 226
Interaction (see a.c.-d.c. system interaction)
Inversion, 18, 31
Ion production, 216

Light guides 134
Lightning surges, 173, 181, 189, 192
Load rejection, 103, 186

Master power controller, 95, 97, 122
Mercury-arc valve, 3
Microprocessor control, 135, 237
Misfire, 159, 188
Monopolar link, 215
Multigroup convertors, 185

Subject index

Multiterminal dc, 224
  fault detection, 227
  switching requirements, 227

Network harmonic impedance
  effect on filter design, 60
  loci in complex plane, 57, 58

On-load tap-changing, 22, 34, 148
Overhead lines, 149
Overlap angle, 32, 41
Overvoltages, on the dc side, 185
  internal, 188

Per-unit quantities, 99, 108
Phase-locked oscillator, 79, 83, 115
Physical models, 167, 175
Power capability of overhead lines, 210
Power control, 91, 96
Power instability, 99, 117
Power factor (see also Displacement factor) 33
Power/frequency control, 116
Power losses, 1, 30, 64, 134, 211, 213
Power modulation, 115, 117, 121
Power reversal, 92, 226
Protection
  differential, 180
  filter, 182
  overcurrent, 175, 178, 180
  valve group, 135, 177
Pulse number, 37, 43, 51

Q (see Filter, quality of and sharpness of tuning)

Radio interference, 142, 147
Reactive power, 33, 105, 121, 232
Rectifier (see also Convertor) 28
Reenergization of dc line, 174
Reliability, 12, 212, 214, 218, 231
Resonance between filters and network, 186
Reversal of power (see Power reversal)
Ripple
  in direct voltage, 17, 37
  injection, 72

Sea return (see Earth return)
Self-saturated reactor, 105
Short-circuit faults, 158, 164, 170, 177
  unsymmetrical, 172, 187

Short Circuit Ratio, 101, 103, 105
Silicon controlled rectifier (see Thyristor)
Smoothing reactor (see DC reactor)
Stability, of ac transmission, 116, 209
Subsynchronous resonance, 120
Surge arrester, 129, 140, 199, 221
Surge diverter, 143, 222
Surge capacitor, 147
Switching surges, 189, 192, 200
Synchronous compensator, 23, 105, 199
System interconnection, 218

Tap changer control, 94
Telecommunications, 96, 97, 227
Telephone influence factor, 53, 59, 67
Telephone interference, 52, 57, 67
Thyristor, 7
  convertor, 127, 196
  module, 128, 137
  station layout, 130, 140, 156
  valve, 127
    cooling, 134
    quadruple, 129
    tests, 136
Thyristor-controlled reactor, 105
Thyristor-switched capacitors, 105
Transformer, convertor, 22, 130, 142, 148
  inrush, 105, 111, 140
  leakage, 22, 24
  saturation effects, 104, 109, 112, 115, 186
  tertiary winding, 24
Transformer connections, 31, 40, 52
  for increased pulse number, 40
Transient overvoltages, 186
Transient stability, 120, 124
Transients, fast (see also Lightning and switching surges) 188, 193, 202
Triple harmonics, 46, 49, 50, 108, 109, 111, 113

Unbalance, 81
  ac voltage, 43
  convertor components, 43, 49

Valve (see Mercury-arc and Thyristor)
Voltage distortion, 53, 60, 79, 106
Voltage regulation, 103
Voltage stability, 105

Weak ac systems, 105, 172, 233